PATER MUNDI;

OR,

DOCTRINE OF EVOLUTION.

BEING IN SUBSTANCE

LECTURES DELIVERED IN VARIOUS COLLEGES AND THEOLOGICAL SEMINARIES.

BY

REV. E. F. BURR, D. D.,

AUTHOR OF "ECCE CŒLUM" AND "AD FIDEM," AND LECTURER ON THE SCIENTIFIC EVIDENCES OF RELIGION, IN AMHERST COLLEGE.

Εἰ μὴ κατεσπαρμένοι ἦσαν οἱ τοιοῦτοι λόγοι ἐν τοῖς πᾶσιν, ὡς ἔπος εἰπεῖν, ἀνθρώποις, οὐδεν ἄν ἔδει τῶν ἐπαμυνόντων λόγων, ὡς εἰσὶ θεοὶ, νῦν δὲ ἀνάγκη. — *Plato.*

If a man sets out to write a book, let him put down only what he knows — I have guesses enough of my own. — *Goethe.*

SECOND SERIES.

BOSTON:

NOYES, HOLMES AND COMPANY,

No. 117 Washington Street.

1873.

RIVERSIDE, CAMBRIDGE:
STEREOTYPED AND PRINTED BY
H. O. HOUGHTON AND COMPANY.

To the

HEAVENLY FATHER,

TO WHOM WE DEDICATE OUR SABBATHS, OUR SANCTUARIES, AND OURSELVES,

These Volumes,

IN ILLUSTRATION OF HIS BEING AND GREATNESS,

ARE REVERENTLY INSCRIBED.

PREFACE.

WHEN a child says, *It broke itself,* we are not much astonished. Such philosophy might be expected from childhood. But what if the same philosophy comes to us from the lips and pens of full grown men who have taken their degrees in science?

In the course of progressive development we have come to — what? Look about you. Here is an idol, with a crowd of half-naked people dancing about it like madmen. There is a slave tilling the field with a sharpened stick. And, sure as you live, yonder is old Democritus — I should know him among a thousand — standing at the mouth of his burrow; unclothed and uncombed and unwashed; a silly leer on his face; whom all the people say to be mad, and who *is* so mad as to say that God and virtue are mere names, as he writes in the sand with dirtiest of fingers the single word, Ἄτομοι! Why, this is not Christendom in the last of the nineteenth century! It is *heathen Greece,* of some thousands of years ago — Greece without the useful arts, without science, without morals — at least Greece in its very childhood as to these chief things. And the child says, *The universe made itself.* Such is Progress.

If the author of this volume could enter again its locked forms he would feel disposed to add —

1. A few striking examples of organisms, at first set down by experts as being clearly of the same species, but afterward found so broadly unlike in their more interior structure as to be unanimously assigned to distinct species, and sometimes even to distinct genera.

2. Some account of that very instructive civil war now raging among evolutionists, in which scarcely a single principle important to their scheme but is loudly called in question by some first-class authority among themselves.

3. A chapter showing in detail how the sort of argument used in favor of evolution might be used, with equal or greater plausibility, in favor of the heathen doctrine of *metempsychosis,* or some other doctrine which nobody now thinks of believing.

As it is, perhaps this book will serve a purpose

till a better comes to us from the pen of M. Thiers. "I must give a pendant to my book on property. I am preparing it — a work against *materialism*. There is no great distance between the enemies of God and the foes of those who possess anything. Materialism is a folly as well as a peril. I am anxious to confound it in the name of science and good sense. For twelve years I have been engaged in this work; during all that time I have been demanding from botany and chemistry and natural history their arguments against the detestable doctrine that leads honest people astray."

LYME, CONN.

CONTENTS.

I. DOCTRINE OF EVOLUTION.

PAGE

1. Nature 9
2. Religious Bearings 10
3. Present Attitude 14
4. Duel with Theism 19

II. AS EXPLAINING NATURE.

1. A Testimony 25
2. General Estimate 26
3. An Objection 28
4. Not to be Accepted if Adequate 30
5. Cannot be Shown Adequate 31
6. Is not Adequate 37
7. Examples of Evolution 37

III. CHIEF DEFENSE.

1. Alleged Examples 59
2. Harmonies with Nature 60
3. Course of Scientific Experience 63

IV. CONFLICT WITH ONTOLOGY.

1. An Illustration 83
2. Combination no Creator 91

PAGE

3. No Equal from Equal 95
4. No Like from Like 97

V. CONFLICT WITH GEOLOGY.

1. Each Age its own Species 105
2. No Perpendicular Chains 109
3. No Horizontal Chains 112
4. Special Chasms 115
5. Objection 127

VI. CONFLICT WITH THE SCIENCE OF PROBABILITIES.

1. Doctrine of Chances 139
2. Broken Chains 141
3. Spontaneous Minims 148
4. Disjecta Membra 152
5. Overlappings of Species 157
6. Improprieties of Structure 161
7. Organic Limits 170
8. Few Types 176
9. Spiritual Properties 182

VII. CONFLICT WITH SOLAR ASTRONOMY.

1. Nebular Hypothesis 197
2. Estimated 199
3. Central Heat 203
4. Chemical Constitution 205
5. Mechanical Relations 209
6. Rotations 216
7. Revolutions 223
8. What Next? 239

VIII. CONFLICT WITH STELLAR ASTRONOMY.

		PAGE
1.	A Principle	246
2.	Visible Systems	248
3.	No Huge Centers	250
4.	Without Certain Graduations	252
5.	Various Planes	256
6.	Eccentric Orbits	258
7.	Different Chemistries	262
8.	A Dream	269

IX. CONFLICT WITH NEBULAR ASTRONOMY.

1.	Nebulæ	267
2.	Examined by Spectroscope	269
3.	Shown to be Fire Mists?	270
4.	Shown not to be	273
5.	What if they are?	285
6.	The Whole Field	291
7.	A Voyage	297
8.	The Whence and the Whither	300

I.

DOCTRINE OF EVOLUTION.

Anaximenes omnes rerum causas infinito aëri dedit.

St. Austin.

Sed quibus ille modis conjectus materiæ
Fundarit cœlum, ac terram, pontique profunda,
Solis, lunæ cursus, ex ordine ponam
Nam certe neque consilio primordia rerum
Ordine se quæque atque sagaci mente locarunt.

Lucretius.

I. Doctrine of Evolution.

1. NATURE 9
2. RELIGIOUS BEARINGS 10
3. PRESENT ATTITUDE 14
4. DUEL WITH THEISM 19

FIRST LECTURE.

DOCTRINE OF EVOLUTION.

THE Doctrine of Evolution — known also as the Law Scheme, and the Development Hypothesis — in its ripest form, is that all things we perceive, including what are called spiritual phenomena, have come from the simplest beginnings, solely by means of such forces and laws as belong to matter. Suppose all matter expanded into one great cloud of atoms. Then these atoms, *by virtue of properties inherent in themselves,* would, in course of time, come together into worlds, into astronomic systems, into the vegetable and animal kingdoms, and even into that great spiritual realm which is the chief wonder and glory of Nature.

I propose to discuss these views at length, because they seem to me the great, and indeed the only possible, assailant of Theism from the side of science.

It is true that not a few persons of great consideration are disposed to think that the Doctrine of Evolution does not really assail our Theism at all. They say it is perfectly consistent with the existence of God, and even with His being the author of Nature. Supposing the nature of matter to be the proximate source of all natural structures and organisms, with their phenomena, the matter itself may have come directly from the hand of a Creator.

This must be admitted. A positive proof of the Law Scheme would do just nothing toward disproving a creating God. At the same time it is true that this scheme is extremely hostile to Theism and evangelical religion generally. One might conjecture as much from its history. It was started by old Greek atheists — Anaximander, Anaxagoras, Democritus, and Epicurus — in the interest of atheism. It was revived and enlarged in the interest of atheism, at the first of this century, by French atheists — Lamarck, St Hilaire, St. Vincent, and La Place. And, up to the present time, most of its leading supporters, the men who have pressed it with most zeal and intelligence, have been widely astray in point of religious belief. They have been materialists, rationalists, free-religionists. They have been deists, atheists, skeptics. They have been active foes of churches, ministries, Bibles. To a man, such gross errorists are now found catching at the Doctrine of Evolution with great eagerness. They scarcely need to be argued with in its behalf. They are ready to take it on sight. At once it becomes their pet philosophy. They dote on it; they put it forward on all occasions; they loudly advertise us that it is desined to be, at no distant day, the destruction of what they are pleased to call *superstition* — meaning Supernaturalism and the Christian Religion. Especially true is this of the "fast and furious" unbelievers in Continental Europe. These men tell us with shining faces that they already see the beginning of the end; that all the sacred traditions are crumbling beneath the ponderous battle-axe of the new scientific giant. "God is dead," say they, "or if not yet dead, He is *dying*." And they blow a trumpet at the news. Whatever doubt others may have as to the real bearing of the Doctrine of Evolution, these men seem to have no doubt at all. While some Christian people look on the speculation with favor, and still more do not as yet see their way clear to reject it (perhaps lest they should repeat the story of Galileo and his persecutors), these men feel, and are glad to feel, that, both in its practical influence and in its logical sequences, it is quite inconsistent with a reasonable faith in the Bible and in God.

And I think their view is correct. The Law Scheme crowds God away till His great orb loses all sensible diameter. It contradicts that whole idea of a personal Divine interference in the affairs of the world, of which our Scriptures are full. Inspiration and miracles and regenerations and even prayers are scornfully cast out by it, as, at best, mere figures of speech. As to the Bible account of the origin of man, of the stage of advancement at which he appeared, of his fall, and of the way in which he is to be restored and saved — this scheme strikes it squarely in the face. Let men say what they will, evolutionism means *materialism ;* and so denies to man moral character, responsibility, personal immortality ; and so denies the chief use of having a God. "And thus," says Hugh Miller, "though the development theory be not atheistic, it is at least practically tantamount to atheism. For, if man be a dying creature, restricted in his existence to the present scene of things, what does it really matter to him, for any one moral purpose, whether there be a God or no? If in reality on the same religious level with the dog, wolf, and fox, that are by nature atheists — a nature most properly coupled with irresponsibility — to what one practical purpose should he know or believe in a God whom he, as certainly as they, is never to meet as his Judge; or why should he square his conduct by the requirements of the moral code, further than a low and convenient expediency may chance to demand?"

Evolutionism also denies that great class of Theistic *evidences* drawn from the admirable natural objects of the universe, and on which faith in all ages has so largely rested. Indeed, it is not too much to say that in effect it suppresses *all* Theistic evidences: for, after I have admitted that the properties of matter itself will account for all we find within the bounds of Nature, what shall hinder a philosopher from saying, "These atoms are just as easily conceived of as being eternal as is an Infinite Mind. The atoms we know to exist, the Mind we do not know to exist. In this case it is unphilosophical to assume the eternity of the unknown, rather than of the known, as an explanation of the facts. One assumption is simpler than two assumptions." No satisfactory answer can be made to this. Accordingly, those scholars who hold to eternal atomic forces and laws which are able of themselves to build up all the various natural structures, are universally atheists. Founded by atheism, claimed by atheism, supported by atheism, used exclusively in the interest of atheism, suppressing without mercy every jot of evidence for the Divine existence, and so making a positive rational faith in God wholly impossible, the Doctrine of Evolution may well be set down as not only a foe to Theism, but a foe of the most thorough-going sort.

And of late it has become a very aggressive and influential foe. Not so influential, indeed, as some of its friends are apt to claim. Listening to these, one might suppose that the entire scientific world had come over to their way of thinking — that the Development Hypothesis is as much accepted science as is the law of gravitation — that none but theologians and the crudest sciolists now think of calling it in question. And has the age really so swept by us? Have matters actually come to that pass that a single curt and casual word of some young Comtist is sufficient to thrust into a corner that old supernaturalism which has reigned supreme through so many ages? Is this hoary doctrine now only worthy of passing mention as a thing which every well-informed man knows to have been thoroughly exploded some time since ; and against which to offer a serious argument at this late day almost calls for an apology ?

Of course all this is abundantly preposterous. It would be ridiculous if it were not so criminal. It belongs to that well-known policy which tries to gain a battle by assuming it to be already gained. The battle is not gained — very far from it — whether we regard evidences or suffrages. Great scholars, and many of them, and most of them, still bow toward the throne of an Almighty Creator, and toward the Cross of His Son, Jesus Christ. There is, perhaps, no country where scholarship is so apt to be unbelieving as Germany ; and it has even been fashionable among evolutionists to claim, in a vague way, that all the German science and culture are in favor of the new views ; but an actual search by one of our most eminent professors among German publications on the Development Hypothesis, discloses the fact that, out of some thirty works issued within a certain time, more than twenty were against the hypothesis, and these as much superior to the others in ability and in the repute of their authors, as they were in number.

Still, it is true that the hypothesis has come to have a very large following and influence, and threatens to have more. It has taken to itself the dress, the airs, the language, and the ideas of our best science. It speaks with the voices, writes with the pens, and persuades with the reputations of well-known scientific men to whose entire scholarly life and labors it is evidently giving shape. So it has managed to come to great notice and influence. It dwells unmolested under the eaves of Christian colleges. It sits honorably in professors' chairs. It is rewarded for its labors by Commencement honors. It is even invited to expound itself in our Theological Seminaries; and to feed itself to the young men who are about to feed the churches. It no longer confines itself to obscure treatises in the dialect of the learned, but tries to popularize itself to the utmost. It stands forth on the bemas of popular lecture-rooms. It drives the pens of widely read authors. It solicits in quarterlies, monthlies, weeklies, dailies. It affects the language of the common people, and even aspires to deal in the speech of the nursery. It has its tracts and its catechisms, and even its pictures. It has its Apostles' Creed, its Westminster Shorter, and even its "Can you tell me, child, who made you?" And the people and the youth are, to an alarming extent, being snared by such means. You can hear this new gospel stammering by the country roadside, in the village grocery, in the blacksmith's shop of the hamlet, and in the sabbathless cabins of the remote school district among the hills. And if the evil is not hourly creeping up, like the plague of frogs, into all our "houses and bed-chambers and kneading-troughs," it is not the fault of many religious newspapers and Christian book-publishers — the latter publishing freely, and probably blindly, popular works all subtly steeped in the new views; and the others as blindly praising these books and their authors in thousands of Christian families all over the land, and even inviting these authors across the seas to diffuse their views everywhere as itinerant lecturers — sounding a trumpet before them to bespeak attention. And why not — seeing that not a few devout and eminent theists and Christians, looking merely at the unquestionable fact that organization by atomic forces and laws is perfectly consistent with the existence of God, and overlooking the equally unquestionable fact that it is perfectly inconsistent with all *evidence* of His existence (especially with that evidence from the things that are made, which the Christian Scriptures say leaves even the heathen without excuse), have been led to tolerate and sometimes to favor so much of the Law Scheme as philosophically draws after it all the rest.

Altogether this scheme has come to great estate. We find it almost everywhere, doubting, insinuating, arguing, dogmatizing, according to circumstances ; sometimes directly affirming atheism, more often quite silent about it, sometimes stoutly and honestly denying it ; but always, I cannot but feel, practically implying it among such beings as make up mankind. No observant theist can fail to see that it is the great intellectual adversary of religion in our times. The rational battle for religion is no longer on the metaphysical field : it is now almost wholly on the field of the natural sciences ; and the champion of unbelief on that field is the Doctrine of Evolution. And a formidable champion it is. As M. Guizot says, "All those who are still Christians and believers in a supernatural life, must become more united against the invasion of materialistic doctrines." There is no other speculation from which so much is to be apprehended ; none equally seductive and dangerous in all the speculating past. And far more emphatically now than when the words were written, more than twenty years ago, by one of the most eminent of modern scientists, can it be said: "The evangelistic churches cannot, in consistency with their character, or with due regard to the interests of their people, slight or overlook a form of error at once exceedingly plausible and consummately dangerous, and which is telling so widely on society, that one can scarcely travel by railway or in a steamboat, or encounter a group of intelligent mechanics, without finding decided trace of its ravages."

The Doctrine of Evolution deserves much attention — especially since it is not merely the only actual, but also the only *possible*, competitor of Theism as an explanation of Nature. We cannot conceive of any other way of accounting for Nature that has any plausibility about it to a thinking and enlightened age. If we may assume that the world will hold fast to at least some respectable fraction of its present intelligence, we may assume that it will never again entertain the idea, either of a creation by chance or of the eternity of existing organic individuals or races. It will always be the Law Scheme or Theism. It always must be. There is no *tertium quid*. The moment one gives up the idea of creating law, he will, of necessity, fill the vacancy with the idea of a creating God. The Law Scheme is the John o' Groat's house to the atheist. It is quite the last ground on which he can stand. Dislodged from this, there is nothing beyond into which he can step forth but that illimitable Theism which washes and encroaches at his very feet. God may be, if the Law Scheme is true; but God must be, if the Scheme is false. While establishing it would prove absolutely nothing against Theism, refuting it establishes Theism in the strongest manner. Hence every blow on this one enemy is really a direct blow in behalf of God; all sound objections that can be stated against the one are so many positive proofs of the other. And they are proofs that can never become obsolete. They meet not only all present atheism, but all atheism that can appear from this time forward. Of course, such universal and immortal arguments are invaluable. One might naturally be somewhat reluctant to spend much time and strength on an argument that may be made useless at any moment by the shifting course of speculation. But such is not the argument against the Development Hypothesis. It cannot be superseded. The age can never get beyond it. And one can afford to expend himself liberally in the effort to remove what is not merely the most noted, plausible, influential, and violent enemy of Theism in our day, but what is its only possible enemy for all ages to come.

I propose to make this large outlay. And I ask particular attention to the fact that, in making it, I do not suspend for a moment the progress of that positive Theistic argument on which I have already considerably advanced. The case being that of a duel, in which the Law Scheme and Theism are the contending parties, you are to count all things which I may be able to show as making fatally against the one, as being so many positive proofs of the other.

II.

AS EXPLAINING NATURE.

Ζῶα δὴ πάντα θνητὰ καὶ φυτά, ὅσα τ' ἐπὶ γῆς ἐκ σπερμάτων καὶ ῥιζῶν φύεται καὶ ὅσα ἄψυχα ἐν γῇ ξυνίσταται — μῶν ἄλλου τινὸς ἢ θεοῦ δημιουργοῦντος φήσομεν ὕστερον ΓΙΓΝΕΣΘΑΙ πρότερον οὐκ ὌΝΤΑ ; — *Plato.*

II. As Explaining Nature.

1. A TESTIMONY 25
2. GENERAL ESTIMATE 26
3. AN OBJECTION 28
4. NOT TO BE ACCEPTED IF ADEQUATE . . . 30
5. CANNOT BE SHOWN ADEQUATE 31
6. IS NOT ADEQUATE 37
7. EXAMPLES OF EVOLUTION 37

SECOND LECTURE.

AS EXPLAINING NATURE.

THE Doctrine of Evolution, as actually held, consists of three parts: *First*, The Nebular Hypothesis, which undertakes to show how worlds and systems of worlds were made in a natural way from a fire mist; *Second*, The Doctrine of Spontaneous Generation, which undertakes to show the natural origin of life and simplest organisms; *Third*, The Doctrine of the Transmutation of Species, which undertakes to show how all the higher sorts of plants and animals came, by a series of natural changes, from one or a few simple species spontaneously produced.

In regard to this last part of the Law Scheme, Agassiz has written, " I wish to enter my earnest protest against the transmutation theory. It is my belief that naturalists are chasing a phantom, in their search after some material gradation among created beings, by which the whole animal kingdom may have been derived by successive development from a single germ, or from a few germs. I confess that there seems to me a repulsive poverty in this material explanation, that is contradicted by the intellectual grandeur of the universe. I insist that this theory is opposed to the processes of Nature as we have been able to apprehend them; that it is contradicted by the facts of Embryology and Paleontology, the former showing us norms of development as distinct and persistent for each group as are the fossil types of each period revealed to us by the latter; and that the experiments on domesticated animals and cultivated plants, on which its adherents base their views, are entirely foreign to the matter in hand."

This strong testimony against a part of the Law Scheme, seems to me not too strong to be borne against the whole. The whole scheme, cosmical and physiological — deriving worlds in their orderly and balanced arrangements from a fiery cloud by such laws as those of heat and gravity, and gradually peopling these worlds with the marvels of vegetable and animal and rational life in the way of spontaneous generation and transmutation of species — this entire scheme seems to me utterly unscientific. I have studied it long. I have read about it much, and thought about it more. I have listened to the arguments of its enemies, and to the arguments of its friends as well. And, altogether, I must say that, while fully allowing the high scientific character of some leading evolutionists, and the real value of many of the facts they have gathered, and even many interesting analogies between their theory and fact, my opinion of that theory is just as unfavorable from the side of science as it is from the side of religion.

It is not science. It is not even a scientific speculation; only a speculation held by some scientific men, and by still more who are no men of science at all. As best drawn out it is very ingenious, very elaborate, very showy, with not a few plausible agreements with Nature — in these respects not unlike many other things which nobody believes in — but after all deserves quite as severe words as Agassiz and Miller and Sedgwick and Brewster and the younger Herschel and many another great student of Nature have spoken against it. That is to say, it is a dream. It is an air-castle. It is a chain of guesses and possibilities and suppositions, holding up a prismatic bubble. It is the mythology of science; defended by such assumptions, fancies, analogies, odds and ends of truth and error dextrously woven together, as can be brought forward in favor of the stories in our classical dictionaries. Do you see yon turreted and battlemented cloud, not without a certain symmetry and likeness to real architecture, but still without solidity and foundation? That is the Law Scheme: save, this cloud is not only without foundation, but *against* foundation; not only without dear friends among the fundamental conceptions of reason and science, but positively at war with them to an extent hardly known in the case of any other speculation that has pushed its way into notoriety. Despite its notoriety, despite some learned and scientific features, despite the support it has had from some men of great scientific attainments, it may well be doubted if ever another hypothesis made so great a figure on so small a capital.

If any think it hardly possible that so many persons of scientific pursuits should take up the Doctrine of Evolution on very slight grounds, I refer them in explanation to two things. The first is that original sin of mankind, the desire to have just as little of God in the universe as possible; and which is ever saying in the hearts of irreligious men, scientific as well as other, "Depart from us, for we desire not the knowledge of Thy ways." And the second is the curious history of science and philosophy.

Has empty speculation never set up claim to be called science, and had a very considerable following? Have learned men never held to any extravagant opinions? Have they never been obliged to take back things which they have most positively asserted and most zealously fought for? Have not even the mathematicians sometimes exchanged characters with the poets, and been as flighty with their differentials and integrals as are the bay-crowned men who make flight a profession? Has Philosophy, so called, always shown itself to be the same as Euclid? Nay, has it not often fathered and mothered such outrageous and preposterous notions as to almost bring its name into contempt among sensible people? In short, is there not very much in the history of what are called science and philosophy to give color to the charge, "When philosophers set out to be foolish there is no folly equal to theirs"? A very mortifying history indeed, although an autobiography—about as mortifying to a scholar as must be to a Jew that history which his nation has written of itself in the Old Testament! Even the annals of the Inductive Sciences will help one to understand how very possible it is for very slender speculations to get audience and friends and apostles in scientific circles.

After looking over the whole field of Evolutionism, it seems to me that the three following propositions are true in regard to it.

I. *Even if the Doctrine of Evolution were shown to be an adequate explanation of Nature, it should not be received as the true explanation.*

II. *It cannot be shown to be adequate.*

III. *On the contrary, it can be shown positively inadequate in many particulars.*

The first of these propositions has been fully treated elsewhere. In the last lecture of the first volume of *Pater Mundi*, I granted, for the sake of argument, that the Law Hypothesis is an adequate explanation of Nature; and then went on to show at length that, even in that case, the true explanation is to be looked for in another quarter. There is another hypothesis, the common Theistic, which also is adequate. And this has greatly the advantage of the other in several respects, which, taken together, are so weighty as to be overwhelmingly decisive in its favor, according to the principles which in other matters uniformly govern the judgments of both practical and scientific men. I mean chiefly its simplicity, its sureness, its salutary character, and its striking accord with what seem to be the primary convictions and traditions of mankind.

But I now withdraw that admission. I no longer accept the Law Scheme as sufficient to explain Nature. On the contrary, I claim that it cannot be accepted as sufficient, if we would consistently hold fast to the principles that lie at the foundation of all our proven knowledge.

And, first, because it cannot be *shown* to be sufficient.

Of course we are not to be called on to accept its sufficiency without evidence. That ancient, established, and most useful doctrine that God is needed to account for the wonders of the stellar, organic, and spiritual universe, cannot reasonably be asked to abdicate the throne and opulent manors of dignity and privilege it has held for ages in these lands, in favor of a quite unsubstantiated claimant. Can the Law Hypothesis be substantiated — at least so far as to show that matter, with a plenty of time allowed it, can form itself into, say, the bodies and souls of the grandest statesmen, philosophers, and saints? I answer in the negative. The thing is intrinsically impossible. It never has been done, despite the greatest efforts ; and, from the nature of the case, it never can be done.

Suppose one should take a bit of albumen, and after passing a charge of electricity through it, should see, not merely a globule within a globule, but atoms hastening to atoms and arranging themselves into a rude organism in which life at once appears. Suppose, further, that, after a careful study of the conditions under which living organisms have best flourished, he should combine and intensify these conditions in a Forcing Establishment; and, introducing into it the new organism, should actually see it pass in the course of a few hours through all the intermediate forms up to a full man.

Nay, let us make a still larger supposition. We will suppose the following bit of romance from the newspapers to be solemnly true — every word of it. "Dr. Meissner has lately shown a very wonderful experiment before the Berlin Academy of Sciences. After years of patient study, he has succeeded in obtaining a certain white powder. This powder was placed in a hollow glass globe about two feet in diameter, from which the air was, as nearly as possible, withdrawn. The globe was then hung from the ceiling at such hight that it could readily be watched by the members of the Academy. Dr. Meissner then violently agitated the powder by shaking the globe with great force. When the powder had become chaotic in its forms, he allowed the globe to hang quietly, and requested the audience to watch it closely. At first all was confusion ; but soon the powder became brilliantly prismatic and a tremendous motion pervaded the mass. A sudden scintillation of the exterior portions succeeded, and a flash of light shot from them toward the center. At the center was then seen, in rapid process of formation, an intensely bright crystal. This crystal began to revolve slowly, and, as it was the only portion of the whole which had at all approached the solid form, the particles of powder began to approach and unite themselves to it. In all directions the effects of attraction were seen ; and, like myriads of scintillating comets, the atoms rushed toward their sun till all had united themselves to it. And now this sun revolved with ever-increasing rapidity, until, as the centrifugal force overcame the centripetal, the ball in whirling threw off ring after ring, and the rings, breaking, rolled up into planets revolving rhythmically around the central sun. Selecting the third planet from this miniature sun, which represented the earth, Dr. Meissner provided the president of the Academy with a powerful magnifying-glass (*very* powerful) and requested him to examine the earth. It was in its azoic age. Not a trace of life could be seen on the barren rocks, none in the lonely seas breaking unimpeded on desolate shores. The paleozoic age came on, and the eye could trace sea-weeds and the earliest vegetation. So the astonished president went through the mesozoic era, and onward as life increased. Vast vegetable forms, mighty ferns, tossing their giant arms in the gale, appeared. Uncouth monsters crept over the land, and swam in the seas. Convulsions rent the earth's crust, and hurried millions of animated creatures to death. Time passed and men appeared, digging roots and ranging the forest. Cities arose, and history — the story of human woe — was represented on this mimic world."

There is an experiment for you! It beats Munchausen — it almost beats Maillet. It throws completely into the shade those famous discoveries in the Moon which Sir John Herschel was once said to have made at the Cape of Good Hope.

It is to be hoped that the next time the German men of science get a chance at that remarkable glass, they will notice, and make a note of, the histories and productions and people of the other planets of our system, and so set at rest several much mooted questions. It is also to be hoped that they will be at pains to preserve a specimen tribe or two of those primitive human savages, with examples of the roots they were seen subsisting on. Perhaps they may, by following down the stream of development very far, be able even to discover the entire Berlin Academy — not the smallest thing in the world — in session, with Meissner himself at their head, astonishing and enlightening them with his wonderful experiment till their faces shine again. Above all, let them preserve us a specimen of *that*. It would be worth preserving.

But, seriously, suppose such experiments were actually made. Would even they decide whether the force by which the original atoms passed through all those changes, and at last came together into an astronomical system, and the highest vegetable and animal forms, was a force *inherent in the matter itself*, or an extraneous Divine force? No force is seen. All that is seen is the moving atom. And all that the experiments show is that *some* invisible force acting according to law, and the conditions of whose action man is able to supply, does the observed wonders. Now the Divine Force, as commonly received, answers this description perfectly. All intelligent theists claim that God is in the habit of acting in this world in fixed ways and under fixed conditions, which can be discovered by men, and largely supplied by them. Indeed, it is essential to the very idea of a perfect ruler that this be so. Otherwise science and profiting by experience would be impossible. So the experiment would prove just nothing at all as to whether forces sufficient to make a man *inhere in matter itself*. It would only prove that such forces exist somewhere. That they are intrinsic properties of the very atoms has not begun to be proved. And yet who expects to get nearer to a proof of the adequacy of the Law Hypothesis than he would be at the close of such experiments as we have supposed? At present we are some stellar intervals short of even that. We shall need to do that impossible thing of going some stellar intervals beyond it, in order to show that matter, out of its own inherent resources, can construct the miracles of organic and spiritual life.

It is enough to prevent our accepting the Law Scheme, that it cannot be positively shown to be an adequate explanation of Nature. But we have much more than this. We can, I think, show that the scheme *is not an adequate explanation*. To this the rest of the argument will be devoted.

Against this view evolutionists bring several considerations. I will notice these first. And, first of all, I will notice the allegation that actual *examples* of a nebular cosmogony, of spontaneous generation, and of naturally transmuted species have been found. Such examples, if they can be furnished, would be very convincing. No better proof that Nature can do a thing, than proof that she has actually done it.

On examining these examples, however, we find that they are not what they profess to be. The cases of spontaneous world-building turn out to be no cases of *spontaneous* world-building at all; at the very utmost, only cases of worlds becoming formed and arranged in accordance with the laws of heat and gravity. The cases of spontaneous generation turn out to be no cases of *spontaneous* generation at all; at the very utmost, only cases of organic life beginning without the presence of any reproductive germ. The cases of transmutation of species by natural causes turn out to be not cases of *such* transmutation at all; at the very utmost, only cases of transmutation under natural conditions. Granting all that can be claimed for these examples, they leave totally undecided, as we have just seen, the question whether the unseen force that first brings the atoms together, say into an organic form, and then varies that form from age to age, is a force belonging to the atoms themselves, or an outward Divine Force subjecting itself to certain constant conditions and laws. I say, granting that the examples are made out, they do not touch the matter in dispute.

But they are not made out. As yet there are no proved cases of worlds formed and arranged into systems in the way of the nebular hypothesis — no proved cases of vegetable and animal births without common parentage — no proved cases of transmuted species. Certainly no one has ever seen a fog become a world; no succession of observers, since astronomy began, have seen it. This is not claimed. It is merely matter of inference, on the part of some, from certain physical principles and certain appearances in the sky. But the justice of this inference is still largely denied by astronomers as profound and eminent as any. And most of those who in some sort favor it do not think of claiming that it has been *proved*, in any proper sense of the word. They express themselves interrogatively. They make a suggestion. They speak of possibilities and plausibilities. They see some things or many things about the hypothesis that look like the truth; that is all. As to positively asserting that there is a single proved case of world-building in the way of natural law, in all the round of the sky, very few ripe astronomical scholars indeed would do such a thing. And with great reason. In another place I hope to make it plain that there is great reason, not only for this caution, but even for emphatically rejecting the nebular hypothesis as being opposed by the latest discoveries.

As to examples of spontaneous generation among plants and animals, something still stronger can be said. History is against them. Case after case of such generation, as confidently put forward as any, has been exploded and fully given up: out of the large number reported to us a few years ago, there is scarcely one that has not been made so untenable by the researches of Ehrenberg, Pasteur, and others, that it has ceased to be spoken of. And at the present time it is admitted by nearly all the European naturalists, and even by nearly all of them who are decided evolutionists, that as yet not a single instance of organization without seed has been made out. "The time may come," say they, "when this will be done." Some seem to expect that it will come soon; the most eminent are satisfied that it has not come yet. For my part I am satisfied that it will never come. The history will go on repeating itself. These men will continue to hear of their *examples*, and will continue to see them fall to pieces under the mallet of careful inquiry; just as it always has been from the days of the *acarus Crossii* down to Bastian's *bacteria*. So I believe. Elsewhere I propose to show that it must be so. At present it is enough to call attention to the fact that most evolutionists have themselves given up their cases of spontaneous generation; and that those few cases which are still clung to by some, are wholly within that misty region of infinitesimals where one easily sees there must be enormous exposure to mistake. It is an abuse of language to say that a single case of spontaneous generation in either worlds or organisms has been proved; while disproof in long succession has, from the beginning, been the order of the day. Standing at the end of so long a line of castaway examples which have made their noise, and had their day, and regularly fallen to pieces, one after the other — will not all friends of the Law Scheme soon lose heart as to what the future will bring them, and conclude to content themselves with that old-fashioned philosophy which finds in GOD the direct source of all the life and mechanics of Nature?

Evolutionists also allege examples of transmuted species, and cite in proof very many accounts of modifications which natural organisms have undergone in various natural ways.

In regard to these accounts, it would perhaps not be altogether unpardonable if one should modestly suggest — especially to men who are apt to criticise so sharply the facts brought forward by supernaturalists — that there is just the slightest shade of doubt as to the correctness of some of them. We like to have our data sound. Facts are great things, but then, one wants to be sure that they are facts. Who shall assure me that some of these eager partisans have always been sufficiently careful to winnow the wheat from the chaff; that in judging of what is true they have always used the good judgment which undoubtedly they possess; that they have never been too hasty in accepting stories which they would be glad to find true (you know we are all poor, weak creatures); in fine, that the random assertions and undue strength of statement and palpable *non-sequiturs* which a sensible man has no trouble in finding thickly scattered over their books, have not more or less tampered with the facts set at the basis of their reasonings? When I am told by a writer that he understands that a gentleman in Sussex has succeeded in modifying that very plastic bird, the rock pigeon, not only into the fan-tailed pigeon and the pouter, but also into a sort of owl, I begin to shrug my shoulders, and to feel somewhat as I suppose some readers of Maillet felt when they read, "At Marseilles the fishermen daily find in their nets, and among their fish, plants of a hundred kinds, with their fruits still upon them. They there find clusters of white and black grapes, peach-trees, pear-trees, prune-trees, apple-trees, and all sorts of flowers."

Great pains have been taken to hunt up, and also to make, cases of extreme variation in plants and animals. Some men think they have found variations actually crossing the boundaries of species. And I have no doubt they have, if we allow them to use the word *species* as loosely as many do. It is a much abused word. What have really been only varieties of the same species have often been erected into so many separate species. No doubt the air lines arbitrarily parceling out such groups have often been varied across organically, and will be again; and that without going very far. If a man makes the mistake of calling the negroes a distinct species from the whites, he will find that Nature makes little of clearing his little wall; even less than Remus did of leaping over the wall of infant Rome. But if he sets up the bounds of species where two groups refuse to mingle their kinds, he will find no example of variation across such bounds. No such examples are agreed on among naturalists.

But many are disposed to claim that, if they cannot point to any such examples, they can at least point to the *equivalents* of these, in variations quite as large and arduous as would be required to pass the extreme members of a species into adjoining species. They tell us of men who not only look and act amazingly like brutes

— say the dog, the sheep, the goose, the donkey — but also men who look and act more like monkeys than they do like the better sort of men. They say that the interval between extreme specimens of men is quite as large and formidable, anatomically and physiologically and intellectually, as that between the lowest human specimen and some apes. And, to help our slow thought, they show us a very mortifying chart of heads and skeletons in which the ape is shaded off by minute differences into the highest man. "Now look at this carefully. Do you not see that really it is further from that head of Newton to this head of Wamba, than it is from Wamba's head to that of yonder chimpanzee? And do you not see just as clearly that it is further from this Caucasian Apollo to that hunchback Æsop, than it is from that Æsop to yon athlete gorilla, who will bend a gun-barrel as if an osier?"

I answer that the actual variation in such cases does *seem*, especially at first view, as if it might be about as large and difficult as that required to transmute the species. But then there is reason to suspect that this seeming may be deceptive; even reason to believe that it is so. Great inner differences are very often found underlying great resemblances, and almost identity of appearance. Two seeds whose shape and size and skin look very much alike, and which, if cut open, would show very much the same interior, on being planted side by side, are found coming to be very different plants. How is this? Of course, in the dim interiors of those seeds there was as wide a difference as there is between the plants into which, under just the same circumstances, they finally grow. We compare two worms: under that superficial likeness lurks as wide a difference as there is between the butterfly and a crawling reptile. We compare two embryos. They are almost precisely alike both to the eye and the glass: but one has the human nature wrapped up in it, and the other the nature of the swine. In both the ape and the man there are large unknown interiors which we cannot compare: and here may lie hid vast dissimilarities, as in a thousand other cases. Just beyond where that apish man has come, on his way to the ape, there may be an invisible structural limit, as impassable as the unseen Bastile wall which stops the night walk of a prisoner: and that such a limit really exists in the cases alleged is strongly suggested, both by the great spiritual gulf between the two, and by the fact that neither history, nor monuments, nor the strata of the globe show the least change in the general type of any existing species. It has always been, so far as we can go back, just the same island we find it to-day — just as widely separated from its neighbors of the same archipelago. And it is found that all the modifications which men, with much pains, bring about in breeds, are invariably lost when the organisms are left to themselves; as they always have been till lately.

But this is but a small part of the answer that should be made to the claim that variations have taken place within a species, as large and difficult as would be needed to reach a neighboring species. To fairly make out such a claim and have it avail for the purpose of the evolutionist, five things must be proved. 1. *The two species compared are really distinct.* Is he quite sure that his argument does not go to show that the ape belongs to the human family — being some degraded prodigal son, who, generations agone, went into a far country, and there lived like swine so long that he has come to apehood on his way to swinehood? 2. *The variation actually observed took place under purely natural forces and laws.*

This also must be proved. The variation has come to us through that dim and wondrous region pointed at by the words *parentage* and *birth* — in fact through a long succession of such regions — and there are some of us who are not disposed to admit that it is as plain as day that the birth of a man from his equal or inferior has nothing supernatural about it. 3. *The range of the actual variation is really, as well as apparently, as large and hard to be made as that needed to transmute the species.* And this, too, must be positively proved. It will not answer to assume it. As facts stand — with such spiritual and historic gulfs between the species compared, and with so many examples of natural objects that look very much alike, even under the microscope, and yet are known to be widely unlike in their more inner parts — how can I feel sure that in those dim interiors of the man and the ape which our knives and glasses cannot lay open to sight, there are not great and radical physical differences? A vast *terra incognita* as yet belongs to each of these beings. Not Africa itself, known only as to its coast-line and a few marginal districts, is more a field for Bruces and Bakers and Livingstones. 4. *There is no bound of an unstructural sort limiting variation to the species.* And this, too, should be proved. After it has been shown that there is no hidden organic limit, it is still a question whether there may not be one of the inorganic sort. It is conceivable that there may be a chemical limit — even one made up wholly of space and time — a wall as invisible and impalpable as that which keeps us from going to the moon. Indeed, Nature is known to have many such shadowy *ne plus ultras,* as invisible and unstructural as the line of perpetual congelation. Such are those that limit the lives of all plants and animals. Such are those that limit their sizes. That which keeps two very unlike species from intermixture does not seem to be structural, but the contrary. It rather seems like that unsubstantial spell with which mediæval romancers were so fond of making business for their heroes, and which no natural forces could break through. "The knight saw nothing: the road seemed quite clear for miles in advance: but he could no more move forward than if a castle wall stood across his path. Then he knew his journey was at an end. He had reached the enchanted district." 5. *There is no limit, structural or unstruct-*

ural, at any point of the long highway by which we follow up the organic races to their rude beginnings. The evolutionist must prove this, too: for it will be of no use to show that a variation has taken place within a species, equal in amount and difficulty to what would be required to carry some members of it into the nearest neighbor-species, if, at some point still further on, the variation can be stopped by some greater barrier. There must be a clear highway down through all the organic territories quite to the *monera,* and further. At least there must be no bridgeless rivers, no insuperable mountains, no unconquerable barricades of *sans culottes* powers of Nature on all that long road. And it does not follow that because a thief can get into a neighbor's house, he can get into every house on the way to Washington. It is conceivable that some species may be more heavily barred against intruders than are others, and that some may be utterly impregnable. Indeed, we know that some species oppose change much more strongly than others. The goose, for example, is known to remain goose with great obstinacy. So with the ass, of every variety. And there is positive reason for thinking that there is a limit to variation *somewhere*, in the fact that as we approach the extremes of any species, the difficulty of a variation rapidly increases — strongly suggesting that at no distant point the difficulty will be insuperable.

When have all these five points been positively proved? So far as I know, the attempt to do it has never been made. And we have ground for believing that, if made, it would not be successful. And yet these are the five points of Calvinism to this argument of the evolutionist — the essence of the whole thing.

But many friends of evolution evidently have the idea that it is by no means necessary to furnish examples of *large* organic variations, in order to prove that species may be naturally transmuted. To them *all* examples are such proof; especially if they are very abundant, relate to the more essential parts of structure, and show change in almost every direction. It seems to them plain that if Nature can make one small organic variation, it can make another like it, and then another; and so may go on adding equals to equals till at last any given organic interval is passed over — even that which parts a man from sea-weed. Accordingly, they fill up their books with detailed cases, from as wide a field as possible, of small variations in plants and animals; chiefly such as are brought about by skillful human effort. And they treat such cases as if they were virtually so many examples of transmuted species.

Of course they are nothing of the sort. Still it is well to inquire at this point, what force such instances of variation really have. Can evolutionists properly infer from them even the possibility of the transmutation of species by natural causes?

To infer from the fact that you can stretch an elastic a little way, that you can go on stretching it indefinitely; that, because you can creep up the mountain slope somewhat, you can go on creeping till you reach the zenith-moon, or even the beetling brow of yon dizzy precipice not twenty yards away; that, because the prisoner of Chillon can pace freely across his scanty cell, he can extend his walk beyond that leaguer of iron and rock that frowns around him, into all Switzerland, and even into all the world, is not exactly the highest style of scientific reasoning. How know these strange logicians that the variable terms in organic beings are not so many prisoners of Chillon! They have some liberty. There is some length to their chain. There is a space within which they can very freely pace up and down. But how do you know that this space is not that of a narrow cell, and that a few short steps will not bring the promenade up squarely against impassable barriers? This much you do know — that natural limits must be reached, sooner or later, on all the observed lines of organic variation. Who supposes that the breed of sheep which can be improved somewhat in size and color and range of diet and fineness of wool and duration of life and length of tail — say into a South Down or a Thibete — could, with any amount of time and pains, be made as large as a man-of-war, or as pictured as a peacock, or as little of a vegetarian as a tiger, or as long-lived as an astral system, or as long-tailed as the most favored comet; or even cease to be unmistakable mutton? Who supposes it? And who knows but that these anatomical, or physiological, or circumstantial dead walls which must at last be reached, may be reached soon; in fact, be reached before that line of species is crossed which the races have never yet been found crossing, and where they resolutely refuse to mingle?

Indeed, the stress of appearance is all toward showing that the variable terms in plants and animals are like those belonging to many mathematical formulæ. In these formulæ the variables are mixed with great controlling constants, and the amount of variation is never such as to alter the specific shape of the whole symbol. Look at those formulæ which express and solve the Higher Astronomical Problems. Here on the earth we have certain organic systems which we call plants and animals. Out yonder in the depths of the sky we find certain other elaborate systems which we call satellite systems, planetary systems, and solar systems These celestial systems have the advantage of being expressible mathematically. Examining their mathematical expressions, we find that in no instance does a variable term vary indefinitely, and in no instance does the sum of all the variations in any formula avail to alter radically its form. Each scheme of worlds holds fast to its specific character, whatever secondary changes it may undergo. The size of each orbit varies, its shape varies, its inclination varies, varies its place in the system; in short, almost everything about a celestial system is changing constantly. And there was a time when it was yet a grave, unanswered question whether such changes might not go on heaping themselves up indefinitely, and at last bring ruin to our Solar System. Some predicted the worst. Why not? The changes were real, they were numerous, they were confessed by all astronomers, they were creeping forward with the steady ease and determination of a natural law; nowhere were there visible goals toward which the perturbations traveled: and not a few trembled and said, There are no goals, and the System must pass on and on to dire confusion and wreck — just as some say now in view of a cooling sun or retarding ether. Then La Grange arose. He invoked the aid of the mightiest Calculus. "Tell me," said he, "what is to be the upshot of this? Must we all perish in the endless ongoings of these celestial variables?" To these unpromising questions he grappled his giant Geometry. It was a hard struggle; every muscle heaped itself into knots, and quivered; but at last the giant overcame. Every change turned out to be periodical. Every variable element reluctantly gave up to view the twin *Ultima Thule* hidden in its bosom. Not only were these limits real, but in most cases they were not far apart. They were also found to inhere in the very Law of Gravity; that is to say, in the very constitution of the variable itself. So our System was safe. And there went up from the whole scientific world a joyful shout as if for a great deliverance. The great Problem of the Stability of the System of the World was crowned. Crowned was the great Geometer. And from that day to this the name of La Grange has been green both summer and winter; and every astronomer who notices a change going on in the sky, under the influence of gravity, at once sets it down as a prisoner of Chillon, able to go only the length of a short chain, and closely shut in by hopeless walls. The heavens are full of such imprisoned variations. Not one of them but is pent within narrow limits by the very constitution and underlying law of the variable. Who shall say that it is not so with these earthly variations which evolutionists make so much of — the variations among plants and animals? It is even easier to believe in "metes and bounds" to these, than it was to believe in them as belonging to the astronomical variations before the discoveries of La Grange. These earthly perturbations are less fluent, less steady, less general, less large, and on the whole less suggestive of unobstructed progress and indefinite continuance, than the perturbations in the sky. Without any Geometry, one seems, at times, as he peers through the glooms, to almost or quite catch the glinting of the chain that cannot be broken, and of the prison wall that cannot be passed; especially when he looks toward the hybrids. In them he thinks he almost *sees* the permanence of species. At all events, as he looks, it becomes easy to believe that through many changes all our fauna and flora maintain substantial identity of type; and that, if they could be expressed by mathematical formulæ, we should find these formulæ as steadfast in their general form as are those which express the history and assure the equilibrium of the Heavenly Bodies.

We do not need to furnish the evolutionist with positive proof of the permanence of species. It is enough to show that it is an altogether unwarrantable and most unphilosophical thing to say, that, because a plant or animal can be varied a little toward the limits of its species, it can be varied victoriously across those limits, and indeed across the whole kingdom of animated Nature. What more credible than that a Creator might see reason to give a certain elasticity to organic beings, to enable them to accommodate themselves to certain changing conditions, and yet see reason to confine that elasticity within certain narrow limits?

III.

CHIEF DEFENSE.

Πῦρ καὶ ὕδωρ καὶ γῆν καὶ ἀέρα, φύσει πάντα εἶναι φασί· -εχνῃ δέ οὐδεν τούτων. — *Plato.*

Ipsa sua per se sponte omnia dis agere expers. Ex omnibus rebus omne genus nasci posset; nil semine egeret. — *Lucretius.*

III. Chief Defense.

1. ALLEGED EXAMPLES 59
2. HARMONIES WITH NATURE 60
3. COURSE OF SCIENTIFIC EXPERIENCE 63

THIRD LECTURE.

CHIEF DEFENSE.

I HAVE considered one part of the argument in support of the Law Scheme. This lies in certain supposed *examples* of natural world-building, of spontaneous generation of organic beings, and of transmutation of species by the forces and laws of matter. It was said that these examples are not real; and that, if real, in the sense meant by evolutionists, they would still leave it unproven that the forces and laws whose results are seen, belong inherently to matter.

I now proceed to two other parts of the general argument of evolutionists. One is from *certain harmonies between their scheme and observed fact;* the other is from *the course of scientific experience.* The first topic I shall speak of very briefly. The other will be treated more at length, as its greater plausibility deserves.

Evolutionists invite us to consider the many

resemblances, more or less striking, between different sorts of animals, and even between animals and plants; also, that obvious gradation in organic beings which enables them to compose a sort of animated staircase, from so simple a thing as seaweed up to so high and complex a thing as man; also, the general advance of the races in grade along the geologic ages; also, the nascent features of some species prophesying of next later species; also, the successive aspects of the human fœtus suggesting strongly the alleged organic progress of the fossils; also, occasional monstrosities and rudimentary organs; and so on. Such facts are admitted. They have been admitted, most of them, from time immemorial. Do these men tell us anything new when they tell us that the body of a man has many points of likeness to an ape, or even to a fish, or even to a worm, or even to a vegetable, or even to the unorganized mineral? We knew that before. It was known long ages before Darwin and his disciples were born. A certain very old book did not neglect to inform the most ancient times that man has a lower nature that allies him to the brutes, as well as a higher nature that allies him to God; and even that this lower nature was made out of the dust of the earth.

Hearing the friends of evolution, one might suppose that the world had not all along been aware of this class of facts ; that evolutionists were the recent discoverers of it : whereas all they have done has been to give some new illustrations of a doctrine as old as the hills, and which, time out of mind, has been doing service in the cause of religion. Yes, most certainly, we admit such facts. And we even admit that they harmonize with the Law Scheme. Doubtless, if the world was made in the way this scheme suggests, we ought to find them.

But what then ? Does it follow that we should not find them all the same, if each distinct species came directly from the hands of a Personal Creator? Who does not see it — all these things agree just as well with the traditional hypothesis of creation as with the other. What is there hard in the idea that God made all the natural organisms with a thread of unity running through them ; made them of different grades ; introduced these different grades at different times, after an orderly fashion, beginning with the lowest ; follows a certain standard process in continuing the succession of each grade ; and so on ? It is true that monstrosities and rudimentary organs, if they could be proved both normal to the system and useless, would be against the *Christian* Theism : but the fact that we do not happen to *see* the use — say of incipient wings, feet, teeth, tails — is certainly no proof that such use does not exist. What accomplished physiologist but will readily confess that his knowledge of organic functions is exceedingly narrow and imperfect? How often have uses been detected which had never been suspected ? In view of the mere history of science, that is a rash philosopher who affirms that the only use of wings is for flying, of feet for walking, of teeth for eating, and of tails for brushing away backbiters.

And, besides, was ever an hypothesis so absurd that it did not find some things, nay, many things, to agree with in this crowded universe ? I would like to give you some exquisite specimens : but I fear all your sobriety would desert you. Suffice it to say, that not the wildest dream in the whole range of the grotesque and absurd and impossible, has failed to touch and coincide with the actual, at some points. It were strange, indeed, if the Law Dream were worse off in this respect than any other. Of course it is perfectly consistent with many facts. But the question is not whether it is consistent with many facts; but whether it is consistent with all facts, and especially with the leading. There is not a crooked stick to be found which will not fit many parts of the broad earth's surface; but how badly it expresses the general shape of the earth, and the underlying glorious arc of the meridian! Doubtless, the Law Scheme fits kindly to some superficial parts of great Nature: a more pertinent fact is that it does not fit kindly to many of Nature's more essential and generic features; that it is positively in conflict with them as expressed in the leading sciences of the day. This I hope to make plain on reaching the positive side of our subject.

I come now to a more plausible argument of the evolutionists; that from *the course of scientific experience.*

Perhaps this argument cannot be more strongly stated than as follows. Once, almost all phenomena were ascribed to the direct action of Deity; the progress of science has ever been to limit this originally vast field of the supernatural, and to enlarge that of natural causes; this course of things has continued so long and carried us so far that now intelligent men almost universally admit, not only that atomic forces and laws are real, but that immense sections of natural objects, and very many wonderful things, such as some chemical compounds and all the marvels of crystallization, are actually produced by them ; and so it is now one of the most natural things in the world, and even a true dictate of experience, to suppose that we can go on much further in the same direction, and that really, it is nothing but our narrowness of faculty and life that prevents our distinctly tracing all wonders to the same natural causation which we admit gives us the wonders of chemistry and crystallization. Besides, why are not the *reasons* on which we admit the purely natural origin of these latter wonders, just as good for admitting the purely natural origin of those others whose names are plants and animals and astronomic systems ; especially in view of the fact that the lowest of organic beings do not differ sensibly in grade from the highest of the inorganic ?

Such is the argument. I have stated it as strongly as possible. So stated, it has at first view a plausible look ; owing to its having a generous outline of fact, certain quiet assumptions colored to imitate fact, and over all a showy vagueness of language which well hides obnoxious particulars. The really empty argument is made to seem sound in very much the same way that yonder bush, with a rag upon it, is made to seem a true human form in the deceitful moonlight. Something that might pass for a human outline is really there; the fancy quietly supplies some convenient additions; and the vague shimmer of the moon ekes out the awful giant. The child's hair stands on end. And yet the whole thing is emptiness. And so is this argument from the course of scientific experience. Its premises are largely assumptions; and, if they were proven, they would not warrant the conclusion drawn from them.

It is not admitted that the difference between the highest inorganic and the lowest organic being is small. It may be small to the sense, but it is not small to the reason. The humblest organic being has the principle of life, the highest inorganic has none of it; the one has the principle of growth, the other has none of it; the one has the principle of reproduction, the other has none of it; the one is a system of organized instruments conspiring to one result, the other is not necessarily any instrument at all. Certainly these are radically very different things, though seeming so much alike. Is this so very strange? The horse that neighed to the canvas-horse of Apelles and got no answer; the bird that pecked at the canvas-grapes of Zeuxis and found no food; still more the Zeuxis himself who put out his hand to lift the canvas-curtain of Parrhasius, and took derisive laughter instead, might have suggested as much.

The best senses are no infallible popes. No Ecumenical has yet been bold enough to say it of them. They are unable to take any note whatever of many gross differences. Here are two seeds. To all our organs the inner matter of the two seems very much one thing, the same white, unorganized farina; and yet they must differ constituently from each other as much as do the tiny flower and the lordly tree into which the same soil will finally develop them. — Here are two eggs. To all our senses the inner substance of the two seems very much one thing — the same yellow yolk and white albumen — and yet really these eggs must differ constituently from each other as much as do the unsightly reptile and the beautiful bird into which the same incubation will finally ripen them. — Here are two stars. To all common observation they seem quite alike — the same radiant eye — and yet one is a mere lighthouse flame a few miles away, while the other is a solid world on which great nations might dwell, to whose golden skirts a family of planets cling, and whose rays are shot at us across abysses which might almost weary the wings of angels.

So do not say that because in certain cases the organic and inorganic are almost perfectly alike to our gross senses, they may not be very unlike in their more interior constitution. They must be. They must differ from each other as do mysterious life and death; indeed, as do whole systems of such stupendous opposites. They must be like two rays coming to us from opposite sides of the same star. Within our sphere these bright lines are practically one; but the deeper we go into space, the further apart are they, and at last they are found apart by the whole breadth of a mighty sun. Such is the final interval that divides the organic from the inorganic. At least, who can show the contrary?

Again, it is not admitted that the *reasons* on which men receive — if they intelligently receive it at all — that atomic forces and laws are the sources of chemical compounds and crystals, are just as good for admitting that they are the sources of the highest natural structures. These reasons are as follows: Atomic forces and laws are known to exist; they seem equal to that low grade of product; it actually seems produced by them; and there is no assignable reason why the seeming does not express the reality. Now this last feature, to say the least, does not belong to the new case. There are assignable positive reasons, and many of them, why we may not admit that matter ever makes itself into organic beings. Some of these have already been given. And many others I propose soon to give from sciences which are the special pets and boasts of unbelievers. It can even be shown, I think, that principles which underlie the whole body of our experimental science positively demand that we ascribe living organic Nature to nothing short of an intelligent author. We must do this or have no science at all. And indeed no reliable business. For all our common affairs are actually conducted on principles which, if applied to religion, would give us a God who is both direct maker and governor of organic Nature. It is a part of my plan to show this.

Further. I do not admit that Deity was once generally supposed the direct author of almost all phenomena. This has never been the popular faith. The true statement would be, that in unenlightened times men were apt to ascribe all events of a very unusual or *startling* character, not otherwise readily explainable, to direct Divine action; for example, such events as earthquakes, volcanic eruptions, eclipses. It is true that this class of events has gradually come, in the advance of knowledge, to be ascribed to natural causes; but it is not so clearly true that the field of the supernatural, as viewed by men, has at all narrowed in consequence. It has narrowed at some points, and enlarged at others. While the waters have encroached on the great continent here, they have retreated yonder. And, on the whole, no ground has been lost. Indeed, I am disposed to claim that much ground has been gained; that the same science which has enlarged before us the field of natural causation, has more than correspondingly enlarged before us the field of the supernatural; that the same science which has explained many things on purely natural principles, has more than made up for this by greatly enlarging the wonderfulness of countless known objects which cannot be so explained, and by discovering countless other objects equally wonderful which were quite unknown to our recent ancestors. Who does not know that the wonderfulness of Nature has increased on us greatly faster than her explainableness?

I say, *who* does not know it? The one has expanded like the astronomical spaces, the other more like the area of geographical discovery. Ten problems rain upon us, to one solution of a problem. And the further we go, the larger and swifter fall the drops, and the more prismatic with beautiful mystery do they show between us and the sun. So that, with all our explanations, Nature is ever getting more high and deep and awful. The more we know, the greater seem the things to be known. The deeper we go into the structure of any natural organism, the more exquisite and bewildering does that structure seem. It is as when we enter some caves. With every step of advance, the higher swells the grotto, the larger and grander range the apartments, and the less impression do our torches make on the deepening and yet superber glooms.

Never was the universe so wonderful to human eyes as it is to-day. The heavens that shone in at the eye of the Hebrew Psalmist were a mere blank, compared with the Newtonian heavens which shine in at our eyes. The terrestrial Nature that went darkling through the Middle Ages was a mere beggar, compared with the Crœsus-Nature that goes with more than oriental pomp and largess along our highways.

See the exquisite refinements of animal and vegetable structure, which the present lancet and microscope display; see the glorious celestial mechanics that blaze in the foci of our present telescopes and mathematics; see the long series of life-epochs which now bestar to us, with their radiant mile-stones, the prodigious track of the geologic ages—such facts as these, and not natural explanations of such things as thunder and lightning, make the leading feature of our present science! These are really the facts whose scepters govern, and whose coronets dazzle our times.

Accordingly, I suppose that to-day the faith of intelligent and devout theists—men *both* intelligent and devout—in the direct Divine production of profuse natural objects, is not only more intense and broad and firm than ever before, but that it relates to a much larger proportion of particulars. Indeed, such persons now almost universally believe that a direct Divine action is mixed up intimately with the production of all events, great and small, that swarm through the daily universe.

And so it happens that all sorts of things are now more freely made subjects of prayer than ever before. We feel far more at liberty than did our fathers, to carry the smallest items of family and personal interest to the ear of Heaven. The microscope has not lightened in vain. Not in vain has that endless revolver, especially for the last generation, been constantly blazing and reporting away at the minims of Nature. It has reported wonders of exquisite littleness; a populous world hanging from the point of a needle. And we have come to feel, more than ever, that nothing is too small for the personal attention and interference of God. If we are less superstitious than the ancients, I trust we have a wider faith. If the miracles of saints and demons are less believed in now than formerly, I trust God's miracles are believed in more than ever.

I would not undertake to say that unbelievers are not at present making more noise than ever before; that this noise is not more than ever couched in the tones and words and formulas of science; that, on this account, it is not creating a greater danger to faith than ever tried any preceding age. All this I sorrowfully believe. At the same time I believe it would be hard to show that, as yet, the proportion of unbelievers in the greatly enlarged class of scientifically informed men has at all increased. Much harder still — let us say impossible — would it be to show that among those men of this class to whom such words as *conscience and duty and virtue are not mere empty names*, faith in God, and in His direct action in Nature, has grown less as science has advanced. I am confident it cannot be shown. But suppose it can. Suppose it true that the field of the supernatural has gradually narrowed with even this class of persons during the very short, and in many respects crude, time which has elapsed since science began. What then? Do not men, on first receiving sight, sometimes see men as trees walking? Is not the faint and unsteady twilight of the morning, especially to eyes just opened from sleep, often fruitful in mistakes? Even truth has its unaccountable ebbs. Even virtue has its surprising backslidings. Even the stars occasionally strangely retrograde, or seem to do so. And why may not Theism, though as true as truth and virtue and the stars, sometimes go strangely backward? It may have done so, and still there be no warrant for the act in the discoveries which the age has made. What has come to be believed in the disturbed and flickering beginnings of science is one thing; what has actually been shown worthy of belief is another. Every philosopher knows, or ought to know, that those men say truly who say that science, with all its achievements, has never yet succeeded in distinctly tracing anything whatever to mere matter as its efficient cause. Indeed, it has not yet been able to show that force ever belongs to mere matter at all. The most it has done in this direction has been to trace phenomena to some force intimately associated with, and conditioned on, certain forms of matter; but that this force comes from the essential nature of matter, instead of coming directly from a Divine source, it does not show. In no single instance has science gone so far. It is speculation, and not science, that pretends to that remote feat.

So much for the unsoundness of the premises in the argument from the course of scientific experience. But what I would lay most stress upon is that the premises, if sound, would not support the conclusion. Does it follow from the fact that an agent does some things, and is gradually found doing more things than was first supposed—does it follow that this agent does *all* things, and especially things of a vastly higher grade than any it has ever actually been found doing? That were a wonderful style of logic. The Hebrews say that Moses had under him many officers to issue all the smaller matters of government, while the greater matters were issued by himself in person. How widely the man would have erred, who, on finding case after case of those secondary agencies, should have allowed himself to conclude that there were no others throughout all the pilgrim host of Israel; that the great lawgiver himself never appeared with his own personal forces in any part of the administration, however exalted and important!

Take another example. A child comes to hear of the first Napoleon. For a time he very naturally imagines that all the things which he finds ascribed to that sovereign in a general way, were done by him personally. By degrees, as his knowledge improves, he becomes aware that many of these things, even some that were quite conspicuous, were proximately done by subordinates — by cabinet ministers, by marshals, by officers of many lower grades, by mere privates. Now if one should bid this child, on the strength of such an experience, leap to the conclusion that Napoleon was a mere cipher; that he did nothing whatever in his own proper person toward the administration of public affairs; that those subalterns of his issued absolutely every matter, up to the greatest and gravest — would it not be a most absurd proceeding? Logic would laugh at such a logician. The facts would laugh at him. Why, Napoleon was a miracle of personal labor. Though doing many things by others, he reserved to himself a certain high grade of agency to which he alone was competent. On this he daily poured out imperial force and genius. With his own hand he drafted the Code Napoleon. With his own hand he diagramed battles and treaties. With his own hand he signed great warrants of pardon, or death, or nobility. Not only did he personally issue all the higher affairs of his empire, but, in point of fact, all those much-doing proxies were vitalized, in what they seemed to do of themselves, by his magnetic intelligence and force that throbbed away perpetually to the very extremities of the monarchy.

Why may it not be so with God? What is to hinder us from supposing that He, too, has His special plane of agency; that, above that plane on which second causes are found fulfilling their mission, there is another which the Lawgiver of lawgivers and the Emperor of emperors has reserved exclusively to Himself, where He works alone, in His own proper person, the surpassing feats of natural mechanics, celestial and terrestrial, and from whence He pours down on all the wheels of Nature the immense volume and gravity and propulsion of His supreme will? I say, what is to hinder? Would it be so very strange if He, too, with His glorious fund of agency, should refuse to be next to eternally idle — if He, too, with His glorious versatility of powers, should choose to have the range of two modes of causation instead of one mode — if He, too, with such a glorious round of empire, should have occasion for things too great to be done by subaltern atoms, or too great to be done sufficiently well by them — if He, too, most important to be known and with glorious claims to admiration and love, should object to being practically lost in an abyss of proxyship; to be hidden at every point behind a tangled thicket, if not a dead wall, of second causes; to be everywhere separated from the thoughts and feelings and realization of His subjects by a chain of sequences stretching across the whole breadth of

Nature and of inexpressible chronologies, that is to say, across that most bewildering interval supposed to lie between yonder fire mist and this fully equipt solar system populous with Newtons and Paradises ; nay, perhaps across an indefinite succession of such monster intervals, each of which might defy the mightiest computing mathematics ?

Would it be so very strange ? I think not. On the contrary, I claim that nothing would be more natural. For the universe's sake, if for no other, God would be likely to disrelish being so thrust into the background of the picture, so dwarfed in the long perspective of mediators, so dimmed and wasted on human sight by innumerable reflections from innumerable planes of causation. He would be likely to disrelish having our thoughts obliged to travel such tiresome and exhausting distances in order to reach Him ; and then, on that bleakest and dimmest outpost of being, lift up faint and bewildered eyes on a Majesty whose chiefest glories are necessarily hid in twilights and clouds to such jaded, benumbed, and almost swooning faculties.

So what sort of logic is it that infers from the fact that God does many things by atomic forces, that He never does anything directly by Himself? Worse inferring could hardly be found. I particularly beg that it may not be called scientific. It is equally against plainest and countless facts, and against the inherent probabilities of the case. Men are everywhere found combining the direct and indirect modes of causation; are everywhere finding it extremely serviceable to do so; are everywhere able to see that doing so is equally suited to their natures and their interests. Else they would be wretched. Else they would wretchedly sacrifice themselves. Else one half would be subtracted from the meaning and usefulness of their lives. And why may it not be so with God? Do not be so unscientific as to assume that He is an exception. If you must assume at all, let your assumption be in accordance with experience, and not in opposition to it. Especially in view of the admitted immeasurable aptitude of a Divine Nature for all modes of causation; of the universal and immemorial tradition that He uses all; and of the obvious moral disadvantage of His propagating Himself on our notice solely through an endless series of ever-weakening undulations — the obvious moral disadvantage of His always dealing with us at arm's length, from more than telescopic distances, from the furthest extremity of a wand, however magical, that crosses the terrible breadth of all our Geologies and Astronomies.

IV.

CONFLICT WITH ONTOLOGY.

Βέλτιον οὖν τὸν μὲν κόσμον ὑπὸ θεοῦ γεγονέναι λέγειν καὶ ᾄδειν. — *Plutarch.*

Quærit Socrates, unde animam arripuerimus, si nulla fuerit in mundo? — *Cicero.*

Καὶ ἔλαττον δὲ ἑαυτοῦ γεννᾷ. — *Plotinus.*

IV. Conflict with Ontology.

1. AN ILLUSTRATION 83
2. COMBINATION NO CREATOR 91
3. NO EQUAL FROM EQUAL 95
4. NO LIKE FROM LIKE 97

FOURTH LECTURE.

CONFLICT WITH ONTOLOGY.

I HAVE now examined the three leading arguments of the friends of evolution. It seems to me that they are very much such arguments as might be brought, with equal or greater propriety, to encourage a belief in the spontaneous origin of the Giant Cities of Bashan.

No living man ever saw those cities being built by human hands. We know of no chain of testimony that can carry us back to such an event. And yet, not a person, not even the evolutionist, doubts that those silent structures all came from the labor and skill of intelligent beings. We would not listen for a moment to any other explanation of them. And yet one could talk against that universal conviction and sure knowledge, almost exactly as we have just heard men talking against the supernatural origin of plants and animals and astronomical systems.

Hear him. "The old traditional notion," says he, "about these cities is altogether at fault. They are, indeed, very remarkable structures — the great rocky blocks are fairly squared and fitted and piled into very architectural forms, as if for human use — and yet my idea is that they really came in a gradual way, one out of another, by the spontaneous action of forces belonging to the atoms which compose them. Do not laugh, but listen. Perhaps you will not think the opinion so very ridiculous, when you have heard my reasons for it. Just look at the countless minute cells (simplest of dwellings) that are constantly being formed in a natural way: at the countless crystals that are ever building themselves up in the primary geometrical figures: at the many natural grottoes, small and great, furnished almost like palaces, with suites of apartments and columns and tables and vases and thrones: at the shapes which the very clouds take in imitation of the more solid castles and cathedrals below: at the rocks and hills in mountainous districts, piling themselves into fortresses, amphitheaters, domes, towers, buttresses, battlements, and almost everything the architect deals in: in fine, at those many living organisms, very small indeed, but more elaborate by far than the best rock-city of Bashan, and which seem to swarm into being of themselves under the careful experiments of naturalists! These things are very suggestive. I regard them as so many *examples* of what unintelligent Nature can do in the way of developing architecture."

Then this ingenious philosopher, warming with his subject, goes on to exclaim: "Now look at those Bashan cities! See the *gradation* among them, and among the structures composing them! Some are vast, complex, carefully wrought. Others are small, simple, and left very much in the rough. Still others are so rude in form and arrangement as to raise the question whether they are structural at all. Between these are many grades, from a palace for a king to a hut for a coney. Notice, further, that the higher grades, apparently, are of *later date* than the others — are less weather-stained, are less sunk in the soil, are nearer the edge of the desert where the forces of Nature seem most active and powerful. Visit Bozrah and Edrei, and see."

"And I wish you to notice, also, *a process of variation* in such structures. To lay no stress on the changes made by time in those ancient rock-cities themselves — for example, in softening their outline, improving their color, wearing off here and there an objectionable feature, sometimes taking completely down parts that disfigure — to lay no stress on these, look at the great changes, which, wholly apart from intelligent agency, sometimes take place in limestone caverns in a few years. Ten years will bring about a change nearly as marvelous as the original glory of that subterranean palace. A new order of architecture appears. The rocky furniture has been changed almost as completely as if that cave were some temple of fashion. Even the shape and size of the apartments have altered. And all without any human help. But if you choose to take a little pains while some processes of crystallization are going forward, you can determine to a great extent the arrangement of the crystals with respect to each other, and even pile them up in about as many different architectural forms as you please — as many as are shown in the various buildings made by man, and almost as easily as we change the symmetrical combinations of the kaleidoscope."

"And you must not overlook the vein of *resemblance* running through all those structures in

Argob. There is everywhere a family likeness. Everywhere stone, everywhere basalt, everywhere squared blocks, everywhere Cyclopean blocks in Cyclopean walls without cement, everywhere the doors and gates and horizontal roofs of rocky slabs. This most perfect city of all looks as much like yon neighbor city which Porter sees with his glass, as a man looks like an ape; and yet is not wholly unlike the unwalled hamlet of a dozen small stone huts that show between. You can, if your eyes are good, see in each edifice some rudimental feature and prophecy of the one next higher: and, if your eyes are not good, you can find many things about those lonely piles that seem to you obscure, useless, and deformed. And I have no doubt that, if you could watch the development of the architectural idea from its simplest beginning in the mind of a child till his mature life, you would find the same succession of stages as is found in those cities, traditionally, but fabulously, ascribed to Moabite giants of four thousand years ago."

"In short, all is just as it would have been if one city had grown out of another, and all out of the basaltic atoms by merely basaltic forces and laws. If you say that no such growth has been observed in Bashan, I answer that it takes a wonderfully long time for Nature to do such work, and that the time during which Bashan has been watched is a mere nothing. If you say that there are gaps in the chain of likeness and sequence that connects these structures, and inquire for the transitional forms which the theory of development supposes, my sufficient answer is, that they have been swallowed up by the vandalisms of unlimited time ; and that the wonder is, not that some links of the chain have been swallowed up, but that any remain. Is not this enough? See you not how easily objections are met, and how strongly my theory of development for Bashan agrees with observed fact?"

But then our ingenious philosopher suddenly remembers that there is something still better to be said before closing the case: and his voice waxes very confident as he begins to tell about *the course of scientific experience.* "Who does not know that in earliest times almost all remarkable things were supposed due to intelligent agency? As knowledge has advanced, more and more of these things have been traced to unintelligent forces and laws. So the experience of the race is ever pressing us toward the point of believing that such forces and laws are the source of those very remarkable cities. I choose to go at once to the point where all must finally come. Bashan was *developed.* Its cities are crystals. Who says that hammer and chisel, and straining muscle of man, set up Kenath and Kerioth and Keires and their threescore fellows? They are stony Law Schemes. They came forth spontaneously from the blind womb of motherly Nature; and were evolved by little and little, through crevice and chasm and cave and geode and cabin, into Cyclopean castles and palaces."

Thus, at length, our ingenious philosopher makes an end. And he looks about on his audience to see what impression his subtle eloquence has made on them. To his amazement he finds every face ablaze with laughter — or with impatience. Nobody deigns him a reply. And his hearers scatter to their homes, as firmly convinced as ever that the Giant Cities of Bashan rose under the hands of contriving beings; and better convinced than ever of the "beauties of scientific speculation."

Of just as little weight are similar arguments when brought in support of the spontaneous origin of those flying cities of the sky which we call astronomical systems, or of those walking cities of flesh and blood which we call men, or of those still greater thinking cities within us, which both walk and fly, and which we call souls.

I come now to the positive side of the argument. I propose to show that the Doctrine of Evolution is inadequate to explain Nature, by showing that it is in conflict with several sciences, and with each of these at several points. We have only to open our ears to such witnesses as Ontology and Geology and Astronomy and the Science of Probabilities, to hear from each many an emphatic denial of the only scheme which in these days tries to explain Nature without a God. And you should bear in mind that, in such a case, each of these denials — each distinct scientific fact found in conflict with the Hypothesis of Evolution — becomes an independent theistic argument. If we find three such ontological facts, and threescore such geological facts, and three hundred such facts astronomical, we have three hundred and threescore and three distinct arguments of scientific authority for the being of a God. Nay, the case is stronger than this. As we add the facts, we multiply the argument.

THE CONFLICT WITH ONTOLOGY.

However lightly one may think of many speculations which profess to sound the depths of Being, and to bring to light its fundamental conditions, it cannot be denied that some such conditions do exist, have become known, and are of so clear and generic a character as to deserve to be called *scientific.* Among these conditions I suppose to be the following.

1. *No being can reproduce itself in kind.*

2. *No being can produce its own equal, much less its superior.*

3. *No mere combination of beings can produce essentially new properties; the properties resulting must be properties, or modifications of properties, already possessed by the constituent beings.*

Let us consider this last principle first. The Law Hypothesis aims to deduce mind from blind matter. It supposes that the ultimate atoms of which the most intelligent men are composed, are quite without thought, will, and feeling. These attributes mysteriously appear as the result of certain combinations of atoms, which in themselves are utterly unconscious, involuntary, and insensible. Such is the assumption. And necessarily. For no scheme of this sort can face scientific men, save with an offer to account for Nature by means of things known to exist, namely, matter with only such properties as our physical sciences recognize matter as possessing. Besides, it will not do to claim, in opposition to consciousness, that the human mind is multiple — a congeries of many separate consciousnesses, intellects, wills, sensibilities. Still less will it do to admit a host of eternal thinkers and souls of even a very low grade. An eternal intelligence were approaching a God too nearly. So an atheistic Law Scheme is under the necessity of getting everything organic and mental out of such atoms as figure in the natural sciences; atoms with only mechanical and chemical and such properties; atoms altogether without such properties as we call mental and spiritual. Our human minds must come from the mere combination of atoms which themselves neither think, nor feel, nor will. To this doctrine the answer is easy. No possible way of combining atoms can generate essentially new properties. The utmost it can do is to modify properties already possessed. It can intensify, abate, neutralize; that is all. Of course it must be so. You cannot get out of things what is not in them. Arrangement cannot by any possibility become a Creator. No chess-playing with positions, mixtures, combinations ; no conjuring with distances, bearings, proportions, attitudes, times, can start into being a property essentially different from any to be found in the constituent atoms.

It is true that chemical combinations are sometimes said to originate new properties. But we do not mean properties essentially new. We only mean something remarkably different in expression from the old properties, though still of the same essential nature, and perfectly conceivable as resulting from the counteractions and coactions of the old among themselves. Thus the traits of common air are perfectly conceivable as resulting from the agreements and antagonisms of oxygen and nitrogen, though in aspect and effects the compound is largely unlike either constituent. So in other cases. Two forces inclined to each other give a diagonal between them—this principle, with its implications, expresses all we find in Chemistry as well as in Mechanics. But such things as thought, feeling, choice, are not sums, differences, diagonals, of the material. They are essentially different from the gravities and attractions, from the mechanical and chemical attributes with which we are familiar. They differ in conception, they differ in the laws which govern them, they differ in effects, they differ in the means by which they are known; they differ according to that overwhelming verdict of mankind in all ages and countries which has always broadly distinguished between body and soul, matter and spirit.

Pray, how do we know that *any* things differ in kind from each other? Are we ever warranted in saying that things are totally unlike? People do not hesitate to believe in differences; they feel confident that such things as extension and color and hardness differ radically from each other; and yet it is quite without warrant that they do so if there is not a great and insuperable chasm between thought and such properties of matter as the natural sciences concern themselves with. For one, I am not yet willing to quit my hold on the very foundations of knowledge. I must still continue to flatter myself that I know *some* things; and, among these, that spiritual and material properties are mutually inconvertible. They cannot be developed or tortured out of each other. As much even can be said of the fundamental properties of matter. How can one get extension out of gravity, or gravity out of color? Much less can one get choice out of either or all of such material attributes. They are essentially and totally different things. Really, the man who does not see this to begin with, will not see it to end with. How can argument help the man who does not perceive the difference between extension and color? As little will it help the man who does not at once perceive the difference between extension and thought. Such differences, if they appear at all, appear as intuitions. At sight we recognize opposite poles of being in the material and spiritual. They have nothing in common; unless such a hopeless chasm between them as divides the stars be reckoned a common possession. And so, no mere combination of atoms, though the choicest and most dynamical of all; no mere play among themselves of deftly arranged chemistries and mechanics, though as subtle and forceful as ever boïled in crucible or thundered from engine, could begin to convert blind matter into an intelligent and voluntary being. There are present no materials out of which to make him.

Observe also that it is an essential part of the Law Hypothesis, that *ordinary parentage* fully explains the continuance of races. Sandwiched in everywhere with the notions of spontaneous generation and transmutation of species, is the quiet and yet — when one comes fairly to think of it — the astounding assumption, that there is not the slightest difficulty in each sort of plant or animal producing the equal of itself. As if even a God could produce his own equal! Outside of the field now being considered, who ever knew a cause make something of the same grade with itself? The beaver is vastly superior to the dwelling it builds, the bee to its cells, the bird to its nest, the spider to its web. Among the various machines made by man, not one but is vastly inferior to his body, though that is largely aided in its work by the contriving mind; and the products of these machines — say the sewing and pin machines — are always vastly inferior to the machines themselves. And, from the nature of the case, it must be so. Reverently be it said, not even Almightiness can make a man that is able to turn out an organism as admirable as himself, or even anything of the same order of admirableness. It is a pure impossibility in the nature of things. The fountain always does and *must* have a higher level than its stream — the producer always does and *must* tower loftily above his product — and human beings neither do nor can make any approach toward producing those wonderful children of theirs who are their equals, and sometimes their superiors. Children are sometimes, both physically and mentally, *greatly* the superiors of their parents. What a poor explanation do their parents give of such offspring as Milton, and Newton, and Pascal? No, the only sufficient explanation of such persons is found, not in the parents who reverently looked up to them from a much lower plane, but in some Being who looked down on them from that vastly higher plane whence even their greatness seemed as the littleness of grasshoppers.

But the Law Hypothesis does more than claim that the organic races produce their peers and even superiors. It claims that these races reproduce *themselves*, in kind; that they originate beings, not only of equal nature, but of precisely the same sort of nature. As if even a God could make a God! As if even Almightiness could make a watch that is able of itself to make another watch, or to do anything toward such a feat! Can any power get four out of two? Suppose an organism composed of a pin machine and a machine for making pin machines. Can such a thing, by any manner of means, reproduce its own sort, or do anything whatever in that direction? The pin machine can turn out pins, and the machine for making pin machines can turn out pin machines — at least with a plenty of aid from watching and manipulating men — but there is absolutely nothing left to do the least thing toward a maker of pin machines; which last is vastly the most intricate and marvelous part of the original organism. For, pins are vastly less wonderful than the machine that makes them, and a pin machine vastly less wonderful than a maker of pin machines. Thus an animal composed of a given organism and a system of means for reproducing the organism in kind, cannot reproduce its whole self, but necessarily leaves unproduced, even in part, what is by far the most surprising part of the whole structure, namely, its reproducing system. And this, whether the original structure act mechanically, or chemically, or in any other way. It cannot do the least thing toward reproducing the perfect like, in kind, of itself; which self is not the organism, nor the system for reproducing that, but the sum of the two. So in no case can a plant or animal reproduce itself.

Plainly, the matter is not helped by supposing two similar organisms to be concerned in the reproduction. That to which neither can contribute the least thing, cannot be made by both. So parents are no sufficient explanation of their offspring. These new beings which are continually appearing about us in immense numbers, and of the highest structural grades — these wonderful human beings, for example — need to be accounted for independently of their fathers and mothers, as much as if they were so many Adams newly sprung on the world without any visible means. Ordinary parentage does absolutely nothing toward accounting for them. And this in whatever way parentage may be supposed to act. Call it chemical, electric, physiological, mechanical, all of these together, it makes no difference. What the argument objects to is the thing to be done, not some particular mode of doing it. It objects to a thing producing its greater, or even its equal, by *any* mode. It objects to a thing begetting its like in any conceivable way of action.

The parental forces, in whatever way acting, are, at best, of only the same order with those generated: in whatever way acting, they can include nothing that tends in the least to produce a new system of reproduction.

So the forces and laws included in parents do nothing toward accounting for their offspring. And so the Law Hypothesis, which offers nothing but parentage in explanation, does nothing toward accounting for them. And it can offer nothing better. Suppose it should say that the generating forces and laws are partly from without the parents, and yet are purely material. Then, I answer that this eternal something must be incalculably superior to its product; in fact, of quite another order of being. But what is the order that rises incalculably above Blaise Pascal? Are there any unintelligent, involuntary forces known to us in the whole round of Nature that can look down, as from the stars, on such a Sublime Soul? Can any blind chemicals do it? Can any blind electricities, or gravities, or compounds of such things do it? What can do it save a vast Personal Being, with oceanic intelligence and will? Where can be the source of such swift streams but above the clouds? Doubtless from above the clouds they come — from higher than thy cloud-capped summits, great Andes, and thy dazzling white crown of eternal snows, O highest Alps! "That which planted the ear shall it not hear, that which formed the eye shall it not see, that which teacheth man knowledge shall not it know ?"

To the Hebrew prophet there was but one answer to such questions as these. He knew no greater absurdity than that of making blind, deaf, and unintelligent things the fathers of mankind. No greater absurdity exists. It is of the same order with that which proposes to get, in a natural way, something out of nothing. Surely, inadequate Law Hypothesis! Full surely, O intoxicate Law Scheme; bouleversing thyself, and then supposing the universe to stand on its apex instead of its base! Most surely, O unnatural Naturalism; impossibly deducing like from like, equals from equals, and even the greater from the less, and even the greatest things in all Nature from that which is next to nothing! As surely as that a God cannot make a God, nothing in the parental economy, or anywhere else on the same level, can be anything more than the conditions, arteries, and tools through which a Great Personal Force from above pours along its mighty reproductive energies: and that whole childhood which perpetually freshens the earth and rejuvenates mankind, must have been begotten from far above the human plane; say from the " circuit of heaven" — why not say from that awful Zenith of which we can assert, and toward which we can wonder, but whither neither sight nor thought can climb? *There* is the spring of these swift human rivers. Thence come our broad Amazons, fruitful Niles, and arrowy Rhones. Thence flows down the Parent-Power upon all the world. True *Protozoön* — infinite, instead of infinitesimal, Alpha of worlds and organisms — intelligent, voluntary, personal, august, cloud-enveloped Summit of all things — we reverently pronounce before Thee that most ancient and venerable of all names, GOD!

V.

CONFLICT WITH GEOLOGY.

Ὅτι ὁ Θεὸς πάντα πεποίηκε τὰ ἐν τῷ κόσμῳ, καὶ αὐτὸν τὸν κόσμον.—*M. Antoninus.*

Res sic quæque suo ritu procedit, et omnes
Fœdere naturæ certo discrimina servant.

Lucretius.

V. Conflict with Geology.

1. EACH AGE ITS OWN SPECIES 105
2. NO PERPENDICULAR CHAINS 109
3. NO HORIZONTAL CHAINS 112
4. SPECIAL CHASMS 115
5. OBJECTION 127

FIFTH LECTURE.

CONFLICT WITH GEOLOGY.

ACCORDING to the Development Scheme, we ought to have in each Geologic Period all the organic species of preceding Periods.

Of course, the *protozoa*, or primitive organic germs, are continually being showered on all parts of each Period; and all the lines of development are always beginning anew. If there is no obstruction to the progress of these lines — if each Period has congenial circumstances for them all, and there is free transit for them all between the Periods — then, of course, each Period will have living in it all the species of the earlier Periods, and will only differ from them in having *some* more advanced organisms. Now, as a matter of fact, each Period had throughout congenial places for receiving these germs and developing them along the several observed lines of variation. That is to say, at any given time the earth has had, somewhere, congenial habitats for all actual species of earlier formations. For example, in our own time, there are somewhere on the globe districts and conditions suited to each known fossil species — places of all sorts as to food, temperature, moisture, air, light; land, water, air; marshes, streams, seas; fresh water, salt water, waters shallow and deep; tropical, temperate, and arctic places; in fine, places where every organism known to the paleontologist could be as much at home as it was in that ancient site where it actually lived and flourished. So of all other Geologic Periods known to us as fossiliferous. There is not one of them throughout which all the earlier species of fauna and flora could not have been thoroughly accommodated. Hence there is but one thing wanted to secure the presence in each Period of all earlier organic species. We need free communication between the Periods. We need full opportunity for all the species to get across those yawning *convulsions and exterminations* which, as some say, separate the different formations. Now, according to the Development Scheme, there have been such opportunities in abundance — broad viaducts of safe transit opening from all the "homes and haunts" of each Period into the matching homes and haunts of the Period below. Most evolutionists are now disposed to claim that the Periods and Eras were not separated by destructive convulsions, but glided quietly into each other: in which case there was infinite opportunity of transit. And, in any case, see the countless sorts of highly advanced and widely differing organisms now on the earth! All these, according to the Development Scheme, came safely across whole Periods and Eras on as many different lines of escape. These lines cut the strata at all points. They pass through all classes of habitats. They cross — those untold crowds of them which belong to the most advanced and widely differing species — the whole breadth of fossil Geology quite down into the Silurian. And they are not mere mathematical lines. They are rather so many wide thoroughfares, so many winding Stygian rivers, visiting all the habitats of all the Periods, ever giving and ever receiving species, and finally drifting down on ever broadening bosoms into our own time crowded specimens of the population of every other. So we ought to see strange sights about us. The old trilobites and saurians ought to plod in our modern marshes. The old asterolepis and zeuglodon and enaliosaur ought to swim in our modern seas. The old pterodactyles and moas ought to fly in our modern air. All those strange forms whose mummied relicts stare at us from the cabinets as souvenirs of dead ages, ought to come out of their cases and incrusting stone, to lead over their lives in new homes and haunts of the nineteenth century as much like their old ones as two pennies are like each other. Do we find those old fossils now living? Not one of them. Not a single Silurian species has come down to us; not a single species of any other Geologic Age. Each Age has species altogether peculiar to itself; and even each of the several Periods of each Age has but few species, if any, in common with adjoining Periods. This could not have been on development principles. It is absolutely incredible that not a single individual of the vast army of fossils should have drifted down to us alive through all the great and swarming aortas of the past. Especially incredible is it that not a single specimen of those hardy molluscan species of the Silurian, which would have found easy home all over the globe in all the Eras, and whose individuals were so amazingly numerous as to make up with their flinty remains whole strata, miles in thickness — I say, it is enormously incredible that not a single specimen of such species should have succeeded in escaping out of its own age by any of those countless tunnels — or, if you please, that immense open prairie, wide as the world — connecting it with all other ages. Even supposing we should, as our researches widen, find a few clear examples of species passed over from preceding formations, it would not be sufficient to save the theory. According to it, we ought to find almost an infinite number of such examples. They ought to *swarm* through the rocks of every Era. All our *present* lands and seas should be *alive* with those strange creatures whose ghostly visages peer at us out of the glooms of the most ancient past.

2. *According to the Development Scheme, each organic individual now living, or that has lived since history began, ought to shade away by insensible structural differences along a continuous line of ancestry into some rude mite of a protozoön.*

This I say in full view of the fact that it is beginning to be fashionable among evolutionists to claim that Nature sometimes takes *leaps*, more or less large, on her lines of development.

It is well known that, at least as a rule, monstrosities among the organic races do not perpetuate themselves; that, at least as a rule, hybridism prevents great leaps from species to species; that almost, if not quite, universally, durable improvements in any specific type are made very slowly, and have not spontaneously taken place, to any appreciable amount, since the dawn of history. So that it is a necessary, as well as accepted, part of the Development Scheme that the organic world has advanced to its present high grades in the most gradual manner. It has been an immeasurable creeping. Each organic thing, of any complexity, has come up to its present place through indefinite ages, and by a series of steps so minute that they deserve to be called differentials. Conceive these series as so many lines drawn downward through the earth, and passing through all the links of ancestry, to their respective *protozoa*. These lines are as innumerable as are the thronging individual plants and animals of every name that have lived on the earth for some thousands of years; and pierce the strata at all points — thick as ever darts stood on the shield of a beset warrior, or as grain-stalks on the valley of the Nile. Of these, a number altogether incomputable express series of organisms whose remains were capable of being preserved in the strata, and

— considering the host of individuals belonging to almost every known species — *must* have been preserved in the strata. Where are these long lines of closely graded fossils? Where are these organic perpendiculars whose close-bound sheaves choke all the bowels of the earth? Some of them, at least, ought to have been met with and recognized in the abundant continuous excavations and explorations which have been made with widely-open eyes. As if to make examination easier for us, the strata are often greatly thrown out of the horizontal, so as to place on the surface the whole breadth of successive formations, and thus enable a traveler in going a few miles with his feet to pass through vast periods of time with his eyes. With what result? Not a single one of these innumerable *scalæ* has been seen. Not a single continuous structural grade, of any length, has been made out; though, according to the Development Theory, they are really as thick in every formation as are autumnal leaves in Vallambrosa. We sometimes find, and doubtless shall continue to find, a plant or animal intermediate in structure to two others of plainly different species, and making the gap between them less than till then it had been supposed: but never yet has the gap been so filled up as to allow one to pass in a known natural way over to the other. I know of no evolutionists who show it. They only show that they have been able to reduce the interval between species somewhat. They might do any amount of this sort of work and not help their case in the least. Who denies that different sorts of plants and animals resemble each other, sometimes very closely? It still remains true that no organic chain long enough to connect two species confessedly different has been brought to light. This could not have happened had evolutionism been true. Researches among the fossils have been too close and extensive.

3. *According to the Development Scheme, eaeh organic individual that has ever lived on the earth ought to shade away laterally, as well as perpendicularly, by minute differences into its protozoön.*

That is, there ought to exist, cotemporaneously with any given organic individual, specimens of all the terms of that closely graded series of ancestors which connects that individual with its root-form.

I have already called your attention to the fact that, according to the new doctrine, *protozoa* like that from which any given organism sprung, have been freely coming into existence and freely advancing ever since. Hence, this organism ought to have cotemporaneously existing on the earth all the preceding grades, away down to the aboriginal germ. The perpendicular series ought to appear also as a horizontal series. The line of closely graded ancestors ought to be perfectly duplicated in a line of closely graded cotemporaries. For example, man ought to find, somewhere among the living things of the world, examples of all the links in that ancestral chain which connects him with his *protozoön.* He ought to shade away laterally through present countries as he shades away perpendicularly through ancient strata. So of every other thing now living. What infinite, infinite lines! Do we find any of them? Do we find *any* of them? Can we shade away a man into a tadpole by judiciously selecting from among living and historic animals? Can we do as much for a single living thing? Enthusiastic observers and travelers are not few. Sea and land and air, all round the world, have been vexed by our curious inquiry. Our Natural Histories are getting to be exceedingly bulky. And yet not a single line of closely graded organisms can be made out from all known living Nature. Not even a considerable fraction of such a line. The best we can do is to piece out a few inches, so to speak, from varieties of the same species: then comes a gap which we cannot bridge by the proper transitional forms. The best series we can make out is but a succession of gaps. Now this could not be if the Development Scheme were correct. With infinite graded series of organisms, in all their integrity, lying along our horizon, right under the eyes of mankind, it is simply incredible that we should not be able to find a single considerable fraction of a single one of them.

And the geologists are no more successful. We should not suppose they would be. What is, hints strongly at what has been. If no continuous organic chain is now living, and if we find no sign of such by going back through the ages of history, the fact pointedly suggests, not merely that we shall not find such among the fossils, but that they did not exist to *be* found. Still let us search. So we leave the bright, warm, vocal homes of the living races, and go down with our torches into those cold and silent mausolea where Nature with impartial hand has laid up the remains of the ancient inhabitants of the world. What do we find? What but that the fossils appear to have been related to each other just as the living and historic races are found related! In all directions the same succession of gaps and partitions. We pass with our hue and cry for the missing links from stratum to stratum, but discover none of them, or not enough of them — neither the *organic horizontals* with which each buried era ought to be crowded, nor the crowded perpendiculars. The infinite fossil grades that pass along the strata are fully as scarce to our finding as those passing through the strata. You see the difficulty is twofold. First, we have infinite sinuous ladders and stairs, so finely graded that an animalcule might walk up them, passing upward from the Silurian ; second, we have these infinite ladders and stairs all fallen, like so many felled trees, across the formations, making for each age a closely-woven stony web, whose warp and woof alike contrive to elude the observation of all careful observers. Let those believe it who can. Geology does not believe it.

5. *According to the Development Scheme, we ought to find no very abrupt occurrence of, especially, the higher forms of organism.*

This scheme being witness, each of these forms must have been reached in the way of numberless delicate transitions from something lower. So there ought to be no organic chasms at all. Especially, there ought to be no great chasms. Still more especially, there ought to be no great chasms just back of organisms of the higher grades; for such organisms are found to have less elasticity and variability of structure than the lower. To such chasms as these last the Development Hypothesis especially objects. It makes oath by itself (for what greater has it to swear by) that they never have occurred — that a very abrupt appearance of high organic life, whether in fauna or flora, has never in a single instance and under any pretense ventured to take place.

Well, what do we find? First, we find not a few *particular organs and organic features* of a very high grade appearing with extreme suddenness, with enormous organic chasms directly back of them, with an entire absence for a long distance just behind them of those flights of infinitesimal steps by which alone they should have made their appearance.

Does any one know a more exquisite organ than the eye? And yet the eye in great complexity and perfection is found in the trilobite, at the very threshold of the fossil world: also in those very microscopic infusoria which some men would have us accept as examples of spontaneous generation. The result was reached by a great leap. For a long distance there were none of those cautious and delicate approaches to an eye, such as military engineers sometimes make to a formidable besieged fortress. There could not have been. The eye of the trilobite abuts hard on a general convulsion incompatible with organic life. Further, it abuts on the Azoic Age, an age without organisms of any sort, save perhaps a few sea-weeds and animalcules. So, all at once, an eye of large size leaped into being across the great gulf that divides it from practical zero. No series of constantly increasing dents in a shell, followed by a series of constantly enlarging and improving holes, and these gradually filled in with humors that slowly ripened into lenses, helped to bridge the immense interval. The whole was cleared at a bound. This is just as impossible on development principles, as it would be for one to mount, without a graduated progress, to that utmost dome of St. Peter's which commands the whole broad Campagna.

And there must have been organs of a still higher order than even eyes, at the very outset of the organic ages. Life, Growth, Reproduction — these things were all there in as perfect examples as can be found to-day. And yet these are the very highest and most wonderful attributes possessed by organic beings: and if, as the new doctrine says, they come of mere organization, the organs producing such wonderful things must be still more wonderful. The cause of a thing must be superior to the thing itself. What a bound have we here! It is passing suddenly from the vale of Chamouni to that utmost Alpine summit which displays all Switzerland and Lombardy; without sloping and spiraling our way up through the usual fifty miles of ascent.

Second, we find *entire organic beings* of high grade appearing suddenly — with great structural chasms just behind them — with no finely graded antecedents by the aid of which they might have crept up to their high places. Huge ferns such as are now nowhere seen; huge pines, stout and lofty as any that dominate Norwegian forests, appeared suddenly — with nothing between them and sea-weed, not even the mosses. Huge cephalopods, with shells twelve or fifteen feet long, and of the very highest mollusk structure, appeared suddenly — with nothing between them and nothing. Huge sharks and ganoids, over twenty feet long and of the very highest type of fish structure — with great organic blanks just behind them — began the Age of Fishes. Huge reptiles, from thirty to sixty feet long and of the very highest reptile structure — with great organic blanks just behind them — began the Age of Reptiles. Huge land-mammals, as the Megatheres and Deinotheres and Mastodons, to some of which our largest modern quadrupeds are mere pigmies; huge sea-mammals, as the Zeuglodons, seventy feet long — all with great organic blanks just back of them — began the Age of Mammals. All of these come upon the scene with extreme abruptness; as if evoked by the stroke of a magician's wand.

Now, the Development Scheme does not object to huge and high-graded organisms, but it does object, and that most strenuously, to their occurring by *huge leaps*. It makes oath that they cannot do so. Lower species of the same group must precede them. They must reach their pinnacle by climbing slowly along finely graduated precursors of less dignity. There can be no great chasm as to size or grade of structure between them and the most similar of preceding organisms. You see how such a notion flies in the face of facts. These fossil giants just mentioned — all of them — crowd up hard against general exterminations. All of them have the next lower species of their respective groups after them in time, or at the most with them; never just before them. A great gulf yawns between them and their nearest kindred of the preceding formation — always as to size, often as to grade of structure, and sometimes as to both. The lower steps of the necessary flight are before the climbers, instead of just behind them. There is a sort of broken stairs to come down on, but none whatever to go up on. And this not in a single instance merely; it is the habit of the Geologic Ages. You see the argument is cumulative. It is not to be supposed that Geology mistakes in so many particulars and on so wide a field.

Besides, these are only the *great* chasms. Geology is full of minor ones, which, though not so striking, are really as inconsistent with any known scheme of evolution as are the others. It is really just as impossible to get half-way up the pyramid of Cheops without a flight of steps as it is to get to the top without it. All chasms that deserve the name abhor the Development Hypothesis as much as Nature abhors a vacuum. "Nature makes no structural leap which she can hold," is *history;* and so is the necessary motto of the new scheme for explaining Nature. According to this, it is just as impossible for the properties and laws of matter to reach a whale or a trout, except along a gentle slope of improving organisms, as it would be to build a cathedral without successive tiers of scaffolding, or from the top downward.

But there is a still greater leap than any of these are commonly supposed to be. I mean *a leap from a protozoön to a man.*

It is common to place man alone at the very head of the scale of organic beings. And it is true that, all things considered, he deserves the place. But he does not deserve it apart from his mental and moral characteristics. Viewed apart from these, as he ought to be—for evolutionists may not ask us to allow them to *assume,* in defiance of the beliefs, traditions, needs, and almost sight of all mankind, that mind is a product of bodily combinations—viewed apart from these and considered as a *mere animal,* man is not more wonderful than many other animals. As a spiritual being he is plain king over all the world.

In some highest specimens — men within whose roomy souls might be described the whole orbit of Neptune — he rises almost unspeakably above all other earthly organisms. But it is only in virtue of his superior *spiritual* traits. The moment we strip him of this superiority, the Samson is shorn of his locks, and the king loses his crown: the moment that speaking face and form of his cease to be informed by a lofty and responsible intelligence, whose regal lightnings flash in glances, words, and actions, he becomes merely a better sort of ape.

And an ape he is, structurally, according to the views of development men. They universally accept the ape as being the next extant link to man in his chain of being; while Huxley and others claim that man differs less from apes than apes differ among themselves. Granting for a moment this most unpalatable doctrine, I ask for some light on the *grade* of this brother of ours. We reckon the grade of a machine in view of two qualities, namely, beauty and efficiency; especially, in view of what it can do. If better than another machine in doing difficult things it is reckoned of a higher grade, though, perhaps, it is the simplest in structure. Judged by this principle, the ape is certainly not a higher structure than that vision of beauty the bird of paradise, or that Bucephalus whose "neck is clothed with thunder," or that behemoth who is "chief of the ways of God," or that leviathan who is "king over all the children of pride." It is not the fairest, strongest, swiftest, hardiest — does not show the greatest variety and excellence of organic feats and qualities. Esthetically considered, physiologically considered, considered even anatomically, that gorilla is not more wonderful, to say the least, than any one of a whole menagerie of animals that might be named, living or fossil.

Is the best monkey that ever chattered in African or Asian woods a nobler animal in structure than yonder eagle, who, with eye fixed unblinkingly on the sun, soars so easily out of our sight on his graceful and powerful pinions, and then cleaves his level way at the rate of two hundred miles an hour; or than that lion, whose kingly voice and lithe strength and mighty bound are the terror of jungles and hamlets? Pray, is our brother ape a more wonderful animal than any of those huge Tertiary mammals; or than that Triassic saurian, still huger, before whose frightful jaws and bulk and strength and eyes of a full foot diameter the Paladins of Charlemagne, and the renowned Cid, and Cœur de Lion, and even the monster-destroying Hercules himself, would have trembled and incontinently run away? Indeed, is the human ape at all more wonderful, considered as a mere organism, than that great placoid fish, the asterolepis, swimming in the very door-way of the Devonian; or than those other placoids lately found swimming in the Silurian seas and on the very utmost coast of discovered life — fishes armed in iridescent and exquisitely carved plate-mail, such as no warring monarch ever wore, or Cellini fashioned; fishes that could outswim the swiftest ship, outsee the sharpest human eye, outdo with force and promptness and endurance of muscle the strongest and most agile human athlete?

Indeed, I verily believe that man, whether ape or not, is not more wonderful, simply as a mechanism, than many of those myriad-eyed insects of the Coal Measures; or than some of those animalcules that must have swarmed to meet the tentacula of larger Silurian animals, now forming whole immense beds of limestone. These living mites of eldest time, these *avant-couriers* of organic magnitude, among which or near which development men look for their *protozoa* — we have the analogues and organic equivalents of these (so the scheme demands) under our microscopes to-day: and find them endowed with mouths, teeth, stomachs, muscles, nerves, eyes; in short, with all the leading human organs. As we gaze we are astonished to see the rich hues, the beautiful forms, the graceful movements, the prodigious reproductiveness, the astonishing delicacy of senses and instincts, the amazing agility and strength of muscle which, if reproduced in a man, would enable him to spring like a whirlwind half round the globe. We have been truly taught that the realms of the microscope are fully as marvelous as those of the telescope. Is not the Lord's Prayer, all perfectly printed beneath a pinhead, quite as remarkable as when printed on a folio page? These pocket editions of Nature, these miniature copies of the *Pater Noster* which our searching lenses show us among the animalcules, I hold, are fully as wonderful as our grosser human bodies, as mere bodies. We are not likely to see the man who can prove the contrary — who, for example, can prove that the wheel animalcule is a less exquisite piece of putting together than a living Apollo Belvidere. It is different from a man ; it wants some things that a man has ; but then it has other things that a man wants : and it would be very hard to show that, on an equitable striking of the balance between the two as wholes, the microcosm is not as admirable as the macrocosm.

So of the eagle, the behemoth, the saurian, the placoid. Each is inferior to man in some respects ; but each is so superior in other respects that we cannot say that, on the whole, its structural grade is not equally high with our own. To all appearance it is. Indeed, a man with only the mental grade of these animals is well known to be a far more helpless creature : in the range and quality of the work he can do, he is below almost all the living tribes. What matters it that his ratio of brain to body is greater than theirs ? We are speaking of *organic* grade ; and that fatty pulp which we call the brain is, to all appearance, one of the least organized parts in the whole body ; and no man is entitled to assume that this appearance is deceptive and really covers a wonderful mechanics — or even chemistry or galvanism — which gives birth to all mental phenomena.

And thus, on each abrupt brink of the successive

Geologic Ages, stands at least the organic equivalent of a man. He stands at the very brink of the Azoic, looking down a precipice as steep and profound as stretches from zenith to nadir. How came he hither? By what stairs did he ascend? That finely graduated stairs is not to be found. It did not exist. The wondrous organism came up from the mighty profound by one great leap. What a leap was that! What leaps were all those that began the various Geologic Eras! It was really the leap from zero to a *Man* — that impossibility of impossibilities, according to the Development Hypothesis.

What answer do the friends of this hypothesis make to such considerations? They plead the *imperfection and uncertainty* of Geology. They tell us how small a part of the strata has been examined, and what great mistakes have sometimes been made in the effort to decipher the Ten Commandments from those tables of stone.

It is very true that large parts of the world have not even been looked on by geologists. It is also true that, in the parts examined by them, by no means every cubic foot of soil and rock has been faithfully dug over and sifted. True — and likely to remain true for some little time yet. But that researches among the formations have not been extensive and thorough enough to bring to light the missing links, if such ever existed, in the supposed continuous chain of organic development — this is not allowed so easily. The strata have been examined enough to find thousands of cases where two closely related species appear by multitudes of specimens, and not a single example of those intermediate forms needed to connect them, and which, if they existed, must have been as numerous and easily preserved as the others. The chances against this are so enormous, that some of the friends of evolution have felt compelled to give up the idea of development by minute changes, and to suppose that it has taken place largely by *leaps*.

But this is mere supposition. Not a case is known in which the whole distance from one species to another has been cleared at a single bound. As much has been confessed by leading evolutionists; although a late attempt has been made to show that such cases have been found in that twilight region of the infinitesimals which is almost darkness itself, and where it is about as easy to stumble as it is to walk. And if the passage from one species to another is supposed to be made by *several* leaps, we have just as much right to demand that the transitional forms appear among the fossils as we have in the case of the other hypothesis. The cases are precisely of the same kind: only the links in the one are much larger than in the other. Besides, an hypothesis that allows a new species to be naturally produced in a very short time — almost flashed on the world — is specially open to the objection that never once in all the long range of human history has a new species been known to arise in this way. All human experience, for thousands of years, is against Nature having come down to us after the manner of a rabbit.

"But is not Geology a very uncertain sort of a thing?" Well: I am prepared to admit that geologists are not infallible. And, while I am about it, I might as well admit that no Vatican Council is even *considering* the question of their infallibility. They have made a great many mistakes. Some of their mistakes have been of the "high and mighty" sort, and will not soon be forgotten — great bubbles of crude and flighty speculation, launched into the air with infinite parade, called worlds and science and philosophy, wondered after a little as they rose gayly over the opened-mouthed crowd, then disappearing; generally bursting as they disappeared. Up to the present time geologists have had to take back not far from a hundred different theories. There is sign that they will have to take back some more. If matters go on as they have done, it will not be long before — what with the deep-sea dredgings and other explorations — there will be great shaking in certain quarters. The very text-books are already having indignation meetings in view of the mutilations preparing for them. And no one dreams of denying that on the outskirts of Geology, as in a degree on the outskirts of every other science, there is a debatable land where the light is weak and the footing insecure; where truth and error with uncertain faces still contend doubtfully for the mastery. Time was when Geology was *all* outskirts. The case is not so bad now; but a vexatious suburb, of large breadth, with its umbras and penumbras, still remains.

So much must be admitted. But then the admission is hardly worth the making. I have not asked you to follow me into that contested borderland of speculation. On the contrary, I have strictly confined myself to the *central* geological region of assured knowledge. If there are any conclusions in Geology which may be relied on, they are those just brought forward as being in conflict with the Doctrine of Evolution. What student of the earth doubts that its past was broken up into many ages — that each of these lacked the specific fauna and flora of earlier times — that neither perpendicular nor horizontal organic chains have as yet been found in any of them — that in multitudes of cases high organisms appeared without leaving sign of such graduated antecedents as the Law Scheme requires?

Besides, these teachings of Geology should be good against evolutionists, if not good against anybody else. These men accept, and appeal to, and heavily lean on, *certain* geological teachings for the support of their scheme. That scheme requires an enormous lapse of time since life began in the world, in order to lift its *protozoa* into men — they very freely go to Geology for that fact. That scheme requires a higher temperature for the earth in remote times than it now has — they very freely go to Geology for that. That scheme requires an advance along the mighty slopes of the past in organic grade of being — they quite freely go to Geology for that; and think they find it when they find a very different matter, namely, an advance in the grade of organic beings. They unhesitatingly take these geological facts and found on them as on so much granite. And yet this granite of theirs is not one jot more reliable than those other facts which we have just been viewing. Both rest on the same grade of research and evidence; and the same style of objecting which is used against the one class is just as good against the other.

"The record has been very imperfectly read." Well: how do you know that further examination of this very imperfectly read record will not track fossil birds, quadrupeds, men even — not the organic *equivalents* of men, but men themselves — down to the very earliest fossiliferous formations; just as some persons are plainly aching to do; and just as some mammals have been tracked down from the Tertiary to the latest, middle, and earliest Secondary; and just as fishes have been tracked down from the Devonian to the latest, (and as some say) middle, and earliest Silurian?

"But the record itself is very imperfect." Well: how do you know that some such causes as have made this record imperfect, and suppressed countless links in all the chains of development, have not also suppressed traces of a vastly swifter rate of rock building in ancient times than we see now — traces which would finish doing what the oceanic dredgings have so astoundingly begun to do, and reduce that venerable geologic eternity which we have so much admired, and during which the races have had ample time for leisurely and drowsily climbing to their present dignity along the easy grade of their fluxional steps, to comparatively very pitiful dimensions?

Certainly, we might talk to evolutionists about their facts very much as some of them do to us about ours. The fact is, their suppositions are like some very tall men; they have only to lie down properly at full length, in order to be wherever they wish to go. They do not need any evidence. But if they would still keep to the time-honored custom of giving a reason for the supposition that is in them, and if they feel inclined to go to Geology in part for that reason, by all means let them deal impartially with facts of equal standing and prestige. Let them accept those on the right hand, and those on the left as well. Surely they will not allow themselves to do so unscientific a thing as to take the science where it suits them, and cast away the same science where it refutes them! Surely they will take the whole rounded science as it stands!

What then? Why, our facts are flatly inconsistent with their scheme for explaining Nature; while their facts fully agree with our scheme. It is nothing against the doctrine of an eternal God that millions on millions of years have come and gone since living beings began on the earth; nothing against the doctrine of a God whose name is Law and Order and Progress, that these living beings have risen in grade as to brain and spiritual qualities as they have moved toward us along that bewildering past. On the contrary, such facts are in embracing harmony with Theism. They say benedictions over it. They put warm, though reverent, lips to its august brow. But those other facts which it has been the object of this lecture to set forth, do nothing of the sort to the Law Scheme. They do just the opposite. They assault the scheme with both hands. And they are not a scanty two or three that join in the assault. They are comprehensive, manifold facts. Those breaks in the continuity of organic life are many — those absences from each era of the species of all preceding eras are countless — those absences of organic perpendiculars and horizontals are countless also — those great chasms just back of the higher grades of organs, organic properties, and complete individuals are many as well as vast. And, altogether, the facts march by battalions. They are the brunt and drift of our accepted Geology.

Some years ago, Professor Sedgewick, one of the most eminent of British geologists, wrote the following words: "Were all the anatomists of the earth against us we should not one jot abate our confidence. For we have examined the old records; but not in cabinets where things of a different age are put side by side, and so viewed, might suggest some glimmering notions of a false historical connection. We have seen them in spots where Nature placed them, and we know their true historical meaning. We have visited in succession the tombs and charnel-houses of these old times, and we took with us the clew spun in the fabric of development; but we found this clew no guide through these ancient labyrinths, and, sorely against our will, we were compelled to snap its thread; and we now dare to affirm with all the confidence of assured truth, that Geology—not seen through the mists of any theory, but taken as a plain succession of monuments and facts—offers one firm cumulative argument against the hypothesis of development."

Is this an antiquated testimony? It is as true to the latest geological facts as to those of twenty years ago. You can read in your own college text-book of to-day, by one who stands in the front rank of living geologists, these weighty words, — "Geology appears to bring us directly before the Creator. It leads to no other solution of the great problem of the creation, whether of kinds of matter or of species of life, than this,

DEUS FECIT."

VI.

CONFLICT WITH THE SCIENCE OF PROBABILITIES.

Νοῦν εἶναι καὶ τοῦ κόσμου καὶ τῆς τάξεως πάσης αἴτιον.

Anaxagoras.

Sed neque centauri fuerunt, neque tempore in ullo
Esse queat duplici natura, et corpore bino,
Ex alienigenis membris compacta potestas.

Lucretius.

VI. Conflict with the Science of Probabilities.

1. DOCTRINE OF CHANCES 139
2. BROKEN CHAINS 141
3. SPONTANEOUS MINIMS 148
4. DISJECTA MEMBRA 152
5. OVERLAPPINGS OF SPECIES 157
6. IMPROPRIETIES OF STRUCTURE 161
7. ORGANIC LIMITS 170
8. FEW TYPES 176
9. SPIRITUAL PROPERTIES 182

SIXTH LECTURE.

CONFLICT WITH THE SCIENCE OF PROBABILITIES.

I ASK your attention, in the present lecture, to another witnessing Science — the Science of Probabilities.

This is really a mathematical branch of knowledge; and it is not altogether easy to render its testimony acceptably into our common forms of thought and speech. I will, however, venture to make the attempt.

The fundamental principle of the Science of Probabilities is, that if there is no assignable reason why a given event should occur in one way rather than in another, then in a large number of cases of such an event, it will occur about equally often in both ways; and the larger the number of cases, the nearer the approach to equality. For example, if a penny is tossed carelessly into the air a million of times, it will fall about as many times on one face as on the other. In any given case of toss-up, one result is, *a priori*, just as likely as the other. No reason can be assigned why this result should appear rather than that. Both results are evidently equally consistent with the nature of things, and with the general tenor of circumstances and experience. So we are quite sure that in a multitude of toss-ups both sorts of results will occur about equally often. On this principle has been built up, with the aid of analytical mathematics, and especially of what is called the Calculus of Probabilities, a great Science which stands proved to us both by the logic of geometry, and the logic of experience. It has met with splendid success in its numerous applications to other Sciences, and to the affairs of actual life. Its conclusions accord wonderfully with observation. It is found perfectly safe to venture on them enormous sums of money; and enormous interests of reputation, social economics, and government. Men are thus venturing every day — in all sorts of insurance companies, in political philosophy, in judicial proceedings, in historical criticism, and especially, in Astronomy and General Physics. Observational Astronomy, and indeed, all the Sciences of observation and experiment, are getting to build greatly on the new Science. It is hard to say what class of intelligent men build on it with the greatest readiness and conviction. Philosophers, statesmen, general scholars, men of affairs — all freely unite in admitting its principles and in doing them the highest possible honor, that of trusting their most valued interests to them. Prominent among these are the leading friends of the Law Hypothesis. These are the men who profess to hope some day, largely by means of the Science of Probabilities, to extend the reign of known law over the whole domain of social, mental, and moral facts. Of course they should be the last persons to find fault with conclusions carefully drawn from a science on which they themselves lean so heavily.

Let us examine some of these conclusions.

1. *If matter is self-organizing — if it has in itself certain properties by sole virtue of which, in more or less of time, the atoms come together into all the organic forms of nature — then we ought to find no chasms, especially no wide chasms, between different sorts of organic beings; but they should be seen shading away into each other in every direction by insensible differences.*

This has already been inferred from certain

physiological and historical considerations. I now infer it independently from the Science of Probabilities. Conceive such transition forms as naturally fill the space between two given organic species. All these middle forms are, intrinsically and circumstantially, just as possible and easy, and so just as credible, as the two extreme forms. In advance of a given construction, whether by a short or long process, the chances are just as good for one of those as for one of these. There is no assignable reason why one should occur rather than the other. Both are equally consistent with the possibilities and facilities and likelihoods of Nature. In fact, it is a pure toss-up which it shall be — as much so as it is which face shall fall uppermost when a penny is carelessly cast into the air. Hence it follows that, in innumerable cases of construction, we ought to have as many examples of each transitional form as of each of the others. It is vastly improbable — millions of chances to one — that there should be millions of examples of the one, and absolutely none at all of the other. So there ought to be no organic gaps about us; none between species, and none between genera or kingdoms. Wherever we take our stand, we ought to see countless long lines of closely-graded organisms stretching away from us in every direction. How poorly this agrees with facts you know.

But the Law Hypothesis, as commonly held, not only supposes that Nature organizes itself, but that it organizes itself in a given way. It supposes that matter first brings itself into certain small and rude organic forms, and then gradually improves these forms through long successions of individuals and ages. Thus man shades away through the strata, by insensible structural differences, toward the worm or some other less promising first ancestor. The succession is that of a line, and not that of a ladder — much less that of a ladder whose rounds are so far apart that no human skill and patience could ever pass a being from one round to another. So of the other organic races. Each has reached its present state through a succession of individuals which shade away into each other by minute structural differences, and so form a continuous line of connection with some rude mite of a *protozoön*.

What says the Science of Probabilities to this? It says that, since each of these intermediate types is just as likely to be present and just as likely to be discovered as are its next neighbors on both sides of it, it follows that in case of a vast number of individuals of each type we ought to find as many of one type as of another ; and it would be vastly improbable — millions of chances to one — that we should find millions of one sort, and absolutely none at all of its neighbors ; still more improbable that this should happen everywhere along that prodigious line of development ; more improbable still that this should happen everywhere on all the myriads of such lines which exist ; in fine, infinitely improbable that we should not find a single considerable fraction of a single one of these long and many lines that pierce at all points the domain of our fossil Geology. And yet Geology finds not a single such fraction.

So much for organic perpendiculars. But the Doctrine of Chances has also something to say of organic horizontals. It says that the different races of plants and animals, coexisting now or in any past time, ought to be found melting into each other as the seasons melt into each other — it ought to be so if the common form of the Law Hypothesis is true. For, since there is, *a priori*, no reason why any given *protozoön* should have a nature or circumstances leading in one direction rather than in another across the field of actual organic life, in the case of an infinite number of *protozoa* all directions would be equally taken, and the total protozoic life would be radiate—would be like a star shooting out its light or its gravity toward all points of the sphere. All gaps would be forestalled; no partialities would be shown to certain forms of organic being existing at any epoch above the equally possible, easy, and credible intermediate forms; at the present time and in every past age, parallel and transverse lines of development, touching and crossing each other everywhere, would give us a seamless web of organisms undistinguishable into such groups as we call species, genera, and so on.

Nothing can be plainer than the immense contradiction of these conclusions to the facts of our own living times. Do the organic races now living run together and confuse all their outlines as do day and night, and as all objects seem to do in the twilight? You know how different this is from the fact. Organic groups do, indeed, sometimes very delicately approach each other, so that one is in doubt whether they ought not to be classed together and called by one specific name. Each species has its varieties which do melt into each other almost as the hours melt into each other. By putting these varieties together we can make a short organic chain, a few tiny links long, — a short organic line, consisting of a few dots that touch each other. But then comes a break — an indisputable chasm for which no occupants can be found by our most careful researches — and then, almost immediately, another break ; and so on, until that long continuous line of organic groups structurally touching each other, which has been so surely promised us, millions of chances to one, turns out to be, instead of a line, a succession of chasms, sometimes of enormous dimensions. The chasms amount to vastly more than the occupied spaces. So it is on all the promised lines — chasms, chasms, hardly anything but chasms. Those between species are generally very marked ; between genera, still more marked. And so the intervals go on widening through families and orders and classes and branches and kingdoms. What a space between the Vertebrates and the Mollusks or Radiates ? What a space between the animal and the vegetable kingdom — between things potential with sensation and volition and intelligence, and things wholly vacant of these wonderful properties ? Above all, what a space between the organic and the inorganic — between things having life, growth, power of reproduction, and things having absolutely nothing of these attributes. This last chasm is a great black gulf across which things look hopelessly at each other, and scarcely interchange intelligible signals. If species are separated as satellites of the same planet are separated, then genera are separated as planets of the same sun, orders as suns of the same cluster, branches as clusters of the same nebula, kingdoms as nebulæ of the same universe.

And the same broken character that is seen in the organic being of our own time, and has been seen in all the times of history, is found prevailing in all the fossil ages. These ages were not so hard on bones and trees as are our phosphate-selling and timber-cutting times. And yet the *hiatus* is regnant among them as among ourselves. Nowhere are the different sorts of plants and animals found gently shading into each other on long unbroken lines, whether horizontal or perpendicular, whether running through successive ages or along the expanse of the same age. No considerable fractions, even, of such lines appear. According to the Science of Probabilities, this is altogether incredible, if the Law Hypothesis is true. There is an infinite balance of chances against it

And this is the same thing as saying that it is infinitely improbable that the Law Hypothesis is true: which is the same thing as saying that it is infinitely probable that organic Nature came from God. The Law Hypothesis is the only rival of our Theism.

2. *If Nature is self-organizing, we ought to see numerous organisms of all sorts and of the larger sizes spontaneously occurring around us.*

It is claimed — and it is convenient to claim — that the instances of spontaneous generation, though many, are always among exceedingly small, if not microscopic, objects. At first thought, this would seem very unlikely in a scheme of mere blind Nature. Pray, why should such a Nature exclusively choose the twilight region of infinitesimals for her creations? And when we come to formally question the Theory of Chances in regard to the matter, we learn that if Nature spontaneously generates organic beings at all, she generates many of them in such size and number and way as to force the fact on the notice of the most careless observer. Indeed, we get so far as to learn that there are millions of chances to one against spontaneous generation being confined to the microscopic and twilight world. Once grant that atoms of matter can of themselves come together into some sort of a living organism, and the way is broadly open to admit that they can come together into any living forms that we see. In going so far, we have gone beyond all the difficulties of the case. These lie altogether in the nature of living organism — not at all in its size or grade or rate of formation. One number of organizing atoms is intrinsically as credible as another. One grade of organizing properties is intrinsically as credible as another grade. One rate of formation is intrinsically just as credible as another rate. The same general sort of properties that suffices to make the lowest organic living thing, will, when merely hightened in degree, suffice to make the highest. And this extra degree, considered as belonging to eternal atoms, is just as conceivable, just as self-consistent, just as consistent with a scheme of such atoms, as another degree. Hence it is just as possible and easy, and so just as credible in the nature of things, for atoms to have natures tending to an elaborate organization as to a rude one, to a large organization as to one that is microscopic, to a swiftly formed organization as to one that ripens through a million of years. There is no assignable reason why one should be produced rather than the other. It is an even chance between them. In advance of a spontaneous organization it would be a pure toss-up what it would be, whether large or small, slowly or swiftly formed, high or low in the scale of organization — as pure a toss-up as when a penny is spun carelessly upward. Accordingly a vast number of such organizations would give us as many spontaneous generations of large, elaborate, and swiftly formed structures as of the opposite sorts. But the latter, according to the Law Hypothesis, are all the while occurring about us.

Therefore, mature trees, cattle, men should all the while come into being around us without any perceptible cause — sometimes as suddenly as they say Pallas started from the head of Jupiter, Venus from the sea-foam, Arabian or Norse palaces under the wands of mighty magicians. In other cases, the process of structure would proceed more slowly, and philosophers could stand and leisurely watch it through all its stages. Clouds of atoms would visibly seek their fellows — bones, muscles, sinews would visibly take shape and size — the largest and most elaborate fauna and flora would be spontaneously built up under our eye, as men seemed to be in the vision of the Hebrew prophet: "A noise, and behold a shaking and the bones came together, bone to his bone; and when I beheld, lo, the sinews and the flesh came upon them and the skin covered them above." In some such way the larger plants and animals would profusely build themselves up in full blaze of day. Instead of occurring always in the dim Debatable Land of microscopic life where nothing is easier than mistake, these spontaneous constructions would as often occur in the very center and focus of our field of observation.

There is no lack of *material* anywhere about us. It is as plentiful as water and common earth. This is the stuff we are made of. The matter that composes all these organic beings is, to an immense extent, loose in the soil; dissolved in the water; diffused through the air; at liberty in impalpable dust and smoke and vapors and gases; moving its atoms freely about among each other in all conceivable ways of approach, contact, and association. Indeed, the very same atoms that once were organic beings make up a large part of the loose surface of the world, and even whole compacted Geologic beds. In cases not a few, we have together in a state of great freedom, all the elements composing a given plant or animal — as when wood or coal is burned, or men decompose by the fires of *autos-da-fé* and of crowded battle-fields.

Do we see this adult sort of spontaneous generation? Have we ever heard of its being seen in all historic time? And yet we should both hear of and personally see it to an immense extent — the Law Hypothesis being true. Millions on millions of large and elaborate structures ought to appear spontaneously, instead of none. They ought to abound as the microscopic spontaneous beings are supposed to abound. Hence, millions to one, the Law Hypothesis is not true. And so our Science affirms again that the only competing hypothesis is infinitely probable — that it is infinitely probable that God is the Author of Nature.

3. *If Nature is self-organizing, we ought to see everywhere about us natural Disjecta Membra — odds and ends of abortive organic beings of all sorts.*

We should see not only complete plants and animals spontaneously formed, but also separate fractions of such beings. For example, we should see scattered about disconnected arms, hands, legs, feet, teeth, hearts, heads, trunks; not the ruins of completed organisms, but what seem like unsuccessful attempts at such organisms. We ought to find such organic fractions in all stages of progress — formed, forming, inchoate — and in great numbers. So says the Science of Probabilities. At first glance, one would say that a scheme of mere Nature would not be likely to confine its spontaneous generations to complete beings. Why should it? Why should it incline to completeness rather than to incompleteness? Is it not intuitively certain that there is nothing in the nature of a scheme of blind eternal atoms requiring it to affect wholes rather than parts? To be sure, the parts might be useless and unable to live. But what of that? Would it be out of character for a blind Nature to make some useless and dead things? And, further, is it not absolutely certain that such wholes as might be affected would be largely counteracted and mutilated while on their way to realization? Nature is full of counteractions. Forces meet and neutralize each other on all hands. Chemistry founds itself on the victories and defeats of contending forces. The harmony and stability in astronomical regions come from the equilibrium of contending forces. All organized bodies, sooner or later, are destroyed in the contest between the forces that favor and the forces that oppose organic life. Hence, in advance of a given organization, there is no assignable reason, either in the nature of things or in the actual working of Nature, why it should be complete rather than incomplete — so a pure toss-up — but rather a reason why it should be partly suppressed by the counteracting forces which are known to exist all around in great profusion and strength. On the whole, the incomplete structure is greatly the more likely of the two. For there are a thousand forms of incompleteness to one of completeness; a thousand fractions to one whole; and each of these fractions is, at least, quite as likely to occur as the whole organism. Consequently, there is a host of chances to one that the organism will be partial. Thus in any single instance of organization. In the innumerable such instances during present and historic time, there would, according to the Science of Probabilities, occur vastly, vastly more fractions than integers. Wholes would be the exception, pieces the rule. It would be a world of seeming organic *débris*. It would be a tremendous miscellany of fragments. No Armageddon of a battle-field would present a more dreadful aspect.

Lift up your eyes. Buds, branches, roots, veins, arms, eyes, brains, lungs, hearts ; all sorts of Disjecta Membra as separate as they are pictured on our anatomical charts ; sometimes embryonic, and sometimes rounded out into the ripeness of a finished organization ; sometimes fresh and dripping with the sap of the algæ, and sometimes fresh and dripping with blood as blue as ever coursed under the white skin of a Plantagenet — see them falling through the air, floating in the water, stirring in the sod ! Here you see the horns of an ox ripening apart from the ox itself, there by itself the hoof of a horse, yonder the proboscis of an elephant, yonder still the feather of an ostrich, and still yonder the brow of a man. Armless hands such as say, *Behold*, in newspapers and on guide boards, or such as is said to have written the doom of Belshazzar ; bodiless heads such as are painted on our canvas ; busts and torsos such as are hewn in our marbles ; fleshless skeletons such as might stand for ancient Time himself; single bones as if from plundered reliquaries and catacombs — all such dreadful spontaneous creations stare at us in our walks, especially from amid the great decompositions of laboratories, conflagrations, cemeteries, and marshes. The Dismal Swamps, the burning Moscows, the plagued Londons, the steaming Père la Chaises, show something more startling than *ignes fatui*. It is as if the museums of Comparative Anatomy had been sacked and scattered by unscrupulous vandals. It is as if the memories of ancient surgeons and tyrants and butchers had been emptied into the objective all around us.

So it would be in case the Law Hypothesis were true. The Science of Probabilities stands voucher. Have we, as travelers or historians, ever become aware of such wonders? Has any Gonzalo, in any land, had occasion to say, in presence of such things and with hair on end, "All torment, trouble, wonder, and amazement inhabit here: some Heavenly Power guide us out of this fearful country!" Has a single well-authenticated example of such fragments of natural creation ever come to the knowledge of mankind? Not a single example. As we have seen, this could never have happened under a scheme of mere Law. The chances are millions to one against it. Hence the Law Scheme is infinitely improbable. That is to say, our Science espouses the Theistic branch of the dilemma, and declares it infinitely probable that GOD is the Author of Nature.

4. *If Nature is self-organizing, we ought to see very many and striking overlappings and dovetailings of the different sorts of plants and animals among themselves.*

We should not only see these different sorts delicately approach each other so as to have no perceptible interval between them, but we should see them superinduced on, and mortised into each other in a great variety of striking ways. For example, the head of one animal would be set on the body of another, the wings of a bird on the body of a quadruped, the legs of a brute on the body of a man — and so on, until no earthly animal could be identified by any one of its members. Cuvier could not tell a lion by his paw, or even a donkey by his ears.

Conceive of a lion with the wings of an eagle — a conception very common to the writings and sculptures of many nations. Is not this griffin as easy to a scheme of mere naturalism, as is either of the animal types contributing to it? Is it not as easily conceivable? Is it not quite as self-consistent? Is it not fully as consistent with a scheme of blind eternal atoms? Pray, why is it not? Who can *show* why it is not? Would any sane man undertake to show cause why blind eternal atoms cannot as well incline to this familiar griffin, or indeed, to any of a thousand similar overlapping constructions which might be mentioned, as to any other? Indeed, it is absolutely certain — as certain as Euclid's axioms — that there is no such cause; that there is absolutely nothing in the idea of blind eternal atoms that is not just as compatible with their having natures tending toward those composite forms imagined by poets and artists, as toward those actually found in Nature. Intuitively, the one sort is just as credible in all respects as the other. One could live as well as the other. The chances for both, as parts of an actual living Nature, are equal. In advance of a given organization — whether by a short or long process, whether by direct spontaneous generation or through *protozoa* and an enormously protracted development — it is a pure toss-up which it will be; as pure a toss-up as when a penny flies aloft from a careless hand. Nay, this statement fails to do justice to the case. The world is full of counteractions. Miscarriages abound like the sands on the sea-shore. And were the atoms to tend constitutionally only to such an organism as does not overlap another, they would, to say the least, be quite as likely as not to miscarry on their way to such organization through the maze of contending currents ; would be quite as likely as not to get misplaced and misjoined among the similar organisms cotemporaneously forming in the same neighborhood. So that, on the whole, the chances are greatly in favor of an overlapping organism. The griffin is far more likely to occur than the lion or the eagle. Hence, in an infinite number of such organizations, there ought to be many more examples of these poetical forms than of others : and the probabilities are millions and even infinites to one against our finding no such forms whatever. But we find none *whatever.* The superposition of one species of plants or animals on another exists only in the vision of poets and allegorists, or in the dead creations of painters and sculptors. No centaur, part man and part horse, gallops on our highways — no mermaid, part woman and part fish, swims in our seas — no minotaur, part man and part ox, roams over our pastures — no cecrops, part man and part serpent, glides among our rocks — no faun nor satyr, part man and part goat, frolics in our glades — no harpy, part woman and part bird; no hippogriff, part horse and part bird, flies in our air — no chimera, part lion and part goat and part dragon, anywhere frightens our children or our men. In our forests we cut no trees that bleed, and weep, and complain with human voices. In our homes, we are happy to know, are no women whose tresses are snakes. Nowhere among the haunts of men appears a man with horns like the Moses of Michael Angelo, or with bird-head like an Egyptian idol, or with horns and hoofs and tail like the medieval Satan. No errant Æneas, nor Perseus, nor Hercules, nor Rinaldo — off on his adventures — finds such things to smite in even outlandish places. Such men have no chance to ply their vocation. They never did have. They have always been out of date. Never a single instance of such grotesque monsters — grotesque and monstrous only because unfound in Nature — has been met with in any land, or in any historic ages, or in any fossiliferous stratum. This would not be so if the Law Hypothesis were true. Myriads of millions to one it would not be so — says the Doctrine of Chances, standing with one foot on irrefragable Geometry and the other on equally irrefragable Experience, and holding up to the noon of our time the vouchers of her splendid successes. Let every corporator in an insurance company hear — hear every person who gets insured — hear every social philosopher and statistician — hear every scientific man who takes the mean of a number of observations, whether in Astronomy or Geology or any other science of observation! Hear that the Law Hypothesis is beyond measure incredible. And let this be the same thing to you as hearing that the only competing doctrine, which announces GOD as the Author of Nature, is bright with immeasurable likelihood.

5. *If Nature is self-organizing, we ought to see about us innumerable improprieties of natural construction — redundant, inadequate, puerile, absurd, and horrid organisms vastly more numerous than those of the opposite character.*

For example, species of animals with more or less legs than they can use; or with necks too short for conveniently reaching the pasture; or with eyes in the feet instead of the head; or with stomachs fit only for grass, while the teeth are carnivorous; or with forms and faces as hideous as sometimes appear in dreams, deliriums, and caricature prints of the day.

Can any claim that such animals cannot be as easily conceived of as any; would not be just as self-consistent, considered as mere structures; would not be just as consistent with that general idea of Nature which underlies the Law Hypothesis? To be sure, some of these might not be able to continue to live: but all of them could *begin* to live, or at least to exist as dead organisms. Pray, why not? Apart from a wise designer, is not an eternal impropriety just as possible and easy to the nature of things, as an eternal propriety? Apart from a wise designer, is there any assignable reason in the wide universe why there cannot be atoms that constitutionally tend to puerile, absurd, and hideous combinations, as well as to combinations that are philosophic, exquisite, and beautiful? Evidently not. The axioms of Geometry are not plainer. In advance of a given construction, it is a pure toss-up what it will be — whether desirable or undesirable, foolish or wise, horrid or attractive — as pure a toss-up as when a penny flies aloft from a careless hand. Nay, this is an understatement of the truth. The chaos of disturbing forces is to be taken into account. Each propriety of structure, on its way to the objective, is liable to all sorts of modifications from the crowds of assailants through which it has to run the gauntlet. Quite as likely as not it will get seriously scarred and mutilated in the attempt. So that, on the whole, the chances are greatly against such an organism as wisdom would choose. And, in countless cases of organization, we should be sure to have far more examples of improprieties than of proprieties ; far more superfluities, deficiencies, follies, and frights of structure than the reverse. We should be sure to have infinite examples of undesirable organisms. It would be infinitely improbable that we should find only a few of them. So testifies the Science of Probabilities.

This on the supposition of the spontaneous generation of large and elaborate organisms. But the argument does not depend on this supposition. Whatever mode Nature may be supposed to take in realizing the proper things will answer just as well for realizing great numbers of the improper. For example, the mode supposed in the Development Hypothesis. It is true that some structures may be supposed so improper and ill-adapted that they could not be perpetuated, or even live; and so could never have belonged to such lines of development as the evolutionist imagines. But there are infinite structures of this class that could have belonged to them — just as easily as any — all the caricature forms, all the redundant forms, all the forms in which certain unfitnesses for "the struggle for life" are compensated, or more than compensated, by certain special fitnesses. For the mere functions of living and strife, forms answering to those of our comic journals are just as eligible as more symmetrical and beautiful forms. Some man with a hundred hands, some hydra with fifty heads, some bison with a score of eyes or ears or horns, would, other things being equal, have special facilities for making his way in the world, and for helping competitors to make their way out of it. And that bison with neck too short to reach the pasture, so that he must kneel whenever he would eat, might have this disadvantage as regards competitors more than fairly offsetted by a tougher skin, a better ear, or a larger muscle than theirs. So that on whatever scheme Nature may be supposed to organize, we ought to see around us *innumerable examples* of such outlandish constructions as have been mentioned. We ought to see far more of these than of others; because there are a thousand conceivable organic improprieties to one such propriety ; and because each propriety, on its way to realization, would have to make its way for millions of years through one continuous tangled wilderness of blind objecting forces.

Now what do we actually find ? Where is the famous Monstrum Horrendum ? Where is the Cyclops with one round eye in the middle of his forehead ? Where is double-faced Janus ? Where is Geryon of the three bodies ? Where Briareus fighting with a hundred hands, or Cerberus barking with fifty heads, or Hydra hissing with nine ? Where is the dragon of St. George — where St. Patrick going about with his head under his arm ? Where is horrid Caliban — where the ogres of fairy tales ? Where are the counterparts of those extreme misshapes which children form with wanton pencil or scissors ; or with which Punch has made us familiar in his immense caricatures ; or which seem to stare and chatter from the ceiling on the victim of delirium tremens ? Imagine a given animal changed as much as possible as to the place and proportion of its various organs — and plainly a vast number of such changes might be made, so as to make an animal as monstrous as ever oppressed our breasts in nightmare, without doing violence to the conditions of a living and capitally struggling organization — who ever fled with bristling hair from such an animal? What geologist ever found the remains of such in the bowels of the rocks? A still harder question. Where is that great crowded limbo of organic improprieties — of superfluities, incongruities, puerilities, and absurdities — which this world, as a whole, ought to be? A *Limbo Fatuorum* — is this the sort of Nature that blazes in the astronomical systems, or waves in our earthly gardens and fields and forests, or swarms in conscious life along the expanses of our sea and air and land? Is this the Nature which our scientists study, and our travelers and historians relate? Has ever pilgrim come to such an outlandish Nature in all his audacious wanderings among the outlands of the earth?

You know how strongly such questions may be answered. *Everybody* knows. "Faith, Sir Alonzo, you need not fear: when we were boys, who would believe that there were mountaineers dew-lapped like bulls, whose throats had hanging at them wallets of flesh; or that there were such men whose heads stood in their breasts?" It is open to the most casual observation, that Nature as a whole is just the antipodes of that inquired for. The actual world is apart, by a whole sky, from that which would have been given by creating Law. It is a system philosophic, exquisite, and beautiful in a very high degree. The further our researches go into the mechanism and physiology of plants and animals, the louder grows the call for admiration. Our experience in this direction has already become so great that we know, by a most commanding induction, that the tenor of all future discoveries will be the same ; and that, could we explore down to the uttermost omegas of natural structure, we should still be bound to say, "Wonderful, wonderful adaptations!" The world is no Patent Office stuffed with all sorts of inventions, good, bad, and indifferent: it is rather a Sydenham of prize specimens where ripest artists may roam as learners, and over which waves in graceful affidavit of approval both the banner of science and the banner of the Empire. It is a superb gallery which has wisely plundered all ages and countries for the choicest works of the Great Masters, and to which all coming ages will go for inspiration and models. From kingdoms down to species and varieties of species, there is not a group of organic beings — I say *group* —

that has a single feature of construction which an artist as such could stigmatize as " improper " — in view of their sphere and line of life. Who ever saw a species of animals -- I say *species* — whose eyes were cubes instead of ellipsoids, opaque instead of transparent, of the same density throughout instead of different densities? Who ever found a species with ears where the air could not reach them, or with necks too short for proper feeding, or with the retina back of the mouth instead of the eye, or with more organs than it could turn to account? Indeed, the same thing can be said of almost every individual of every species. No fault can be found. No improvement can be suggested. We do not see how we can either prune or graft the being to advantage — in view of its line of life. In all that bundle of myriad machinery there is nothing which a careful man sees his way clear to pronounce absurd or weak or unwise. Such an economy of means, such short cuts to results, and yet such entire sufficiency — it puts to shame the best machine of human devising.

And it is only at very great intervals that we find things that suggest a difficulty. It is true that we often find things the use of which we do not know: but this was something to be expected. It is true that sometimes we come across in mature individuals rudimental organs which *seem* to be doing nothing of the work which *seems* to be the *chief* business of the organs when mature—for example, incipient teeth, tails, wings—but who knows that the chief business of such organs is their only business, and that in their incipient state they may not answer excellently well some obscure lesser ends in the complex organism? It is true that once in a great while we stumble on a *lusus naturæ*, a malformation and even mon strosity: but such cases are extremely rare for so crowded a world; are plain thwartings of the constructive forces; are plain results of disease or of disordered parental structure, and are soon eliminated from the system. They are an admonition. They signify that natural organisms can get out of order. They proclaim that the parental economy must not be trifled with. *Rari nantes in gurgite vasto*—they proclaim a stormy latitude, and advise all mariners to prudence. In short, they are such things as may well be supposed to occur occasionally in a system created and presided over by One who works by general methods, who sees it best to have His creatures to a certain extent mutually dependent, who has moral as well as natural uses to promote, and who will have the whole earthly system bear some marks of sympathy with and adaptation to that moral derangement which has crept into it.

Such is actual Nature. An incredible Nature to have come in the way of self-organization — says the Science of Probabilities, crowned with the chaplet of her mathematics and splendid successes. The witness says true. The chances are unspeakably against such a system from such a source. That is, it is infinitely likely that the Law Hypothesis is false. And this is the same thing as saying that it is infinitely likely that the alternative Theism is true, and that this wondrous Cosmos of exquisite organisms which so rejoicingly sympathizes with the idea of a God is the actual work of His marvelous hands.

6. *If Nature is self-organizing, we ought to find men, or any other organic group, confined to no one standard of adult size, or of period of growth, or of length of life.*

The individuals of each species should vary among themselves almost without limit in these respects.

Men have an adult stature of about five feet.

They reach this stature in about twenty years. They die in about seventy years. Other species have other standards. Each has its own full size, its own rate of growth, its own age-term. The cedars are old after many centuries, the mushroom after a few hours. The standard bulk of elephants is so much, of sheep so much, of rabbits so much. Ravens get their growth in ten years, horses in four, may-flies in a few hours. And so on. While different species differ greatly among themselves in these respects, the members of the same species differ but little. They follow a common law — somewhat elastic, indeed, but still one.

Now, according to the Law Hypothesis, this ought not to be. Individuals of the same species ought to have the freedom of all the sizes, of all the growth-periods, of all the age-terms — at least within such limits as are actually found in Nature; or rather, within such limits as recognize on the one hand the immense divisibility of matter, and on the other hand the immense abundance of free organic material of all sorts about us. Why not? Why cannot a pigeon incline to live a hundred years as well as a raven or a turtle? What hinders a pilot-fish from being constructed on the scale of the whale? What earthly reason can be given why such atoms as make up human bodies could not as well have natures tending to one bulk as to another? Who can tell why fifty-foot men or five-inch men are not every way as credible, in a purely natural system, as men of five feet? Who would try to tell why such a system should incline to a man who is full grown at twenty years, and old at threescore and ten, rather than to one who is full grown at four, and old at sixteen? I think no one who has ever seen a balance hesitating at equilibrium. It is plain as noon — plain as the noon of *Geometry* — that a very great range of sizes, and growth-periods, and life-terms is equally open throughout to a self-organizing human body: that they all are equally possible and easy to it; all as easy to conceive of, as self-consistent, and as consistent with the nature of a scheme of blind eternal atoms. And this, in whatever way Nature may be supposed to organize men — whether by immediate generation, or by first generating an organic germ and then developing it through countless ages and grades into the best Caucasian. Evidently, nothing depends on the mode. If an adult eye can be developed in all sizes, from that of a monad to the Grecian shield of a deinothere, so may an adult man.

So, *a priori*, it is an even chance what sort of an organism we get as to the three particulars mentioned. In advance of a given organization, it is a pure toss-up what it will be — whether mountain or mote, whether momentary or millennial — as pure a toss-up as when a penny leaps upward from a wanton hand. Who knows what face will fall uppermost? But this all know, that in a *million* of such random casts, one face will appear about as many times as the other. And so we may know that, in case of millions of even-chanced human bodies, one stature, one age-term, one growth-period will not appear more than another. So of any other species of natural organisms that has a great many individuals.

What a world it would be! All around us, as well as in the strata, the fables of the classics and the classic fables of our childhood would cease to be fables. Behold Liliputians, fairies, elves! Behold Brobdignags, Cyclops, and Titans! Lo, little men, for a regiment of whom the palm of your hand would be ample parade ground! Lo, great men, one of whom could place Pelion on Ossa; or, Atlas like, seem to bear up the African heavens! Not merely giants and dwarfs, not merelv Anakim and Tom Thumbs, but veritable monsters as large as the Genii of the Arabian Nights on the one hand; and, on the other, merest specks of humanity as small as those organic points which one has to put into the focus of a powerful magnifier in order to see! Men who can look down on the tallest Sequoia Gigantea of California; and, among them, men who can look up at the shortest blade of grass that ever rose on Arctic plain! Men who, like a fungus, have ripened to full stature in a night; and, among them, men who, like some British oaks, have been growing ever since their fathers came over with the Conqueror, or at least ever since the famous three brothers landed in this country! People of three centuries as well preserved as the Seven Sleepers of Ephesus; Methuselahs as fresh as yesterday; Wandering Jews still as strong to journey as they were in the days of the Cross; and yet, among them, patriarchs bearded like pards and covered with venerable snows, though they have not yet kept their first birthday! Nay, men who, like the yew of Hedsor and the cypress of Chapultepec and the Baobab of Africa, still stand firmly under the weight of their three thousand and five thousand years; and yet, among them, men who, like some insects, are borne quite from birth to old age by the short flight of a summer's morning!

So of other things. Expect next summer mosquitoes as large as eagles, and fleas as large as elephants. Count on seeing in the garden something like Jack's famous bean stalk; or in the poultry yards eggs that tell of Sindbad the sailor. Wonder not at finding the morning-glories durable as those century-glories, the oaks; and your corn as tall as the majestic palms of Palmyra. In short, the even handed lottery has been sowing in the skies, the fields, and the seas.

Such is the Nature we ought to see, in case we are dealing with nothing but Nature. Millions to one we should see it — protests the Theory of Probabilities. It protests by its unimpeachable standing as a science, and by all its triumphantly successful applications to affairs as well as to science, that it is just as incredible that we should not find a single instance in which a species throughout has the range of all the standards in these three respects — and where is there such an instance — as it is that in innumerable random casts of a die it will always fall on the same side. That is to say it is infinitely incredible. Hence it is infinitely improbable that the Law Hypothesis is true; and that anything short of God selected and firmly maintains for each organic species those remarkable standards to which we have seen each individual is full surely, though somewhat elastically, held. Such a disposing of the lot when cast into the lap must have been from the LORD. He it was, who, probabilities beyond measure, determined for each group the times before appointed; set to each the bounds which it cannot pass; said to each — whether tiny moss or mighty conifer, whether momentary mote or enduring man — in regard to size and growth and life, "Thus far shalt thou come and no farther."

7. *If Nature is self-organizing, we ought to find no one plan of structure — indeed, no few plans, but an indefinite number of them.*

As a matter of fact, all natural organisms have the property we call life, and sustain themselves and grow by appropriating matter from without by means of certain organs. This means one common plan of structure for all. Such a plan, the Law Scheme, when consulted by the Science of Probabilities, strongly objects to. For, there are other plans of structure which are just as possible and easy to the world of fact as is this one which involves life and growth. Witness the many sorts of machines which man makes! Witness the vastly larger variety which he can and will make in course of the coming centuries! These neither live nor grow, and are none the less organisms for that. Hence, in the immense field of natural organizations, it is infinitely unlikely that they would all be of this one living and self-enlarging pattern. Multitudes of them ought to be of the nature of watches, pin-machines, cotton-gins, steam-engines. Where are they? Who sees a saw-mill, or anything like it, in process of being formed without hands? When have any such things as spinning-jennies all ready to hum, and ships all ready to sail — or even the beginnings of them — managed to appear without the toilsome contrivance and labor of intelligent workmen? Who finds such things as even the simplest tools of the farmer, the carpenter, the shoemaker, raining from the air; or compounding in the seas, and the sod, and the smoke; or slowly developed along the ages? The man who should wait for the supply of his tool-chest till Nature, of its own accord, came to his aid would, I ween, wait long enough. Whatever else it may be, full surely the Nature we wot of is no spontaneous Trip-Hammer for the manufacture of anything like those lifeless and ungrowing tools, machines, and engines with which human ingenuity has made and armed our civilization.

The one general plan of structure in Nature is found divided into four sub-plans, called the Radiate, Mollusk, Articulate, and Vertebrate. Why *four?* Are there no more plans conceivable and possible for living and growing beings? So far from it, we can conceive of an indefinite number of them just as realizable in their nature. Witness, again, the countless sorts of human inventions not belonging to either of the four natural classes; though they are as easily conceived of as having vitality and growth as are woody fibre and the bones of animals. Should living wheels, velocipedes, chariots roll along our streets; should living hoes, plows, reapers toil in our fields; should living chairs, tables, pianos rap and leap and sing in our houses, as we are told they sometimes do for the Spiritualist — why, for the life of him, no one could give a reason why such things are not fully as much in harmony with that notion of Nature that belongs to the Law Hypothesis as are any animals that we see. When one reads in the Hebrew prophet of those wheels full of eyes in which was the spirit of the living creature, and which came and went like a flash of lightning

— while he recognizes the novelty of the conception, he does not recognize it as being a conception of the impossible. Just as possible to the world of fact as is a man! Consequently, in a prodigious throng of living and growing constructions, it is infinitely unlikely that they should all turn out to be of only four patterns — as incredible as it would be that a castaway penny should fall on the same face only four times in a million of throws.

Again, each of these four sub-plans is found divided into a *few* others. For example, the vertebrate plan appears under the forms of fishes, reptiles, birds, and mammals. These forms are only four out of indefinite millions equally possible and credible. You might sit down and with wanton pencil prolifically design, for a lifetime, vertebrates which would be neither fishes nor reptiles nor birds nor mammals, and yet be fully as fair candidates as they for a place in a merely natural scheme of a living and growing world. Why not romance at Natural History as easily and plausibly and plentifully as at any other sort of History?

Further, each of the four vertebrate plans is found divided into a *few* others. For example, the reptile plan appears under the forms of turtles, lizards, and serpents. Are these three the only conceivable reptilian forms? Say three billions rather — just as possible and credible, for aught one can see, as the forms that actually exist — say as many as would affright us were all the uncouth monsters ever stealthily drawn by the unfledged and lunatic pencils of children, small or great, on slate and fly-leaf and desk and disfigured wall, to suddenly step forth into life, cold-blooded, oviparous, and scaly.

And so on — down as far as naturalists have distinguished and classified structural differences. It is a long succession of independent selections. And then this selection rapidly spreads out like an open fan as we go downward, until, at last, it covers the whole earth with some hundreds of thousands of collateral selections for as many different collateral species. Under each group are preserved, at most, only a few types of structure, out of indefinite multitudes equally credible in a scheme of mere Nature. Almost all the sands on the sea-shore fall through the coarse web of the tempestuous sieve: it is only here and there a grain of special largeness that remains on the reluctant wires. Now this is altogether contrary to the Doctrine of Chances. In case of toss-ups almost without number, it is incredible that the penny should always come down on the same face, save in four instances, more or less. Perfectly incredible, I say: as indeed is the idea that Nature has only *one* plan of spontaneous generation (if any), namely, that of rude and microscopic germs in one or four, or at most but few, types; only *one* plan of growth for the individual, namely, that by self action for a certain small proportion of its whole life; only *three* modes of animal reproduction, namely, the oviparous, the viviparous, and the polypal; in short, only *one* or a *few* plans of almost anything, from the one law of gravity, downward or upward. And you observe that, altogether, what we have is an incredibility of an incredibility to the *n*th power. The question is on the joint occurrence of many independent selections, each of which is supremely incredible by itself. The problem is one in geometrical progression. What think you of a differential of the millionth order? It is the wonderful denominator of such a fraction as this which scientifically expresses the unlikelihood — why not say impossibility — of the joint occurrence of this long Chinese alphabet and gamut of incredible things. So says the Calculus of Probabilities. See the huge way in which this Science, with all its well-earned laurels and stars and crosses of the Legion of Honor, pronounces against the Law Hypothesis and for the Theistic! Beyond measure and beyond measure — says this solid witness — it is sure that there is a *Person* who carried through into the objective this long quadrant of selections. No lottery, though bearing the grand name of Law, would have done it. We must look toward a still grander name — one that dazzles as we look. With bated breath, call it God. He it was who chose here one, there four, yonder two, and almost everywhere a few, plans of structure and process out of thronging multitudes that might as easily have been taken; and then swept the mighty remainder into the limbo of rejected ideas where the curious and leisurely may still find them. It is thus we have that unity of Nature in which philosophy rejoices, and in which religion sees mirrored the august face of One Eternal Creator.

8. *If Nature is self-organizing, the following things should be true of Mind. It should be without freedom, without moral character, without just accountability, without power of being influenced by*

perceived motives, without immortality or even a future beyond this world, without limitation of degree, and without limitation of the higher degrees to men or even to organic forms.

The Law Hypothesis proposes to account for all natural things, including mental phenomena, by means of the admitted inherent properties and laws of matter. It supposes that Mind is solely the result of the coming together in certain ways of certain blind material atoms. Of course it must be as inflexible and necessary, in its mode of action, as the matter from which it solely springs is shown to be by the whole tenor of our experience and of physical science. The child must be like the parent. The stream must be like the fountain, however long and tortuous the flow. We must think and will and feel by *fate*. There can be no more freedom in the workings of our minds than there is in gravity or chemical actions. This is admitted, and even claimed, by most friends of the Law Hypothesis. We are mere organized stones — say, if you prefer, organized heat or light or electricity.

Of course, such things as virtue and vice, and just responsibility for anything they are or do, are impossible to men. No one has been justly punished or blamed since the world began. — Of course, also, men cannot be influenced by perceived motives of any degree. Our views, choices, feelings, voluntary conduct, can only be modified by modifying the physical causes on which only they depend, namely, the unperceiving atoms, or that unperceiving arrangement of the atoms which incubates them into intelligence. All arguments and eloquence, all examples and appeals to reason or conscience or interest, all the boundless talkings of men with their fellows to get them to do or not do certain things, amount to nothing. They are altogether nugatory and absurd. The only way to alter the opinions, purposes, and feelings of men is to alter the number or the arrangement of the particles on which such things solely depend. While the cause remains the same, the effect must not be expected to vary. — Of course, further, the human mind is not immortal. We are brothers to the brutes. Come on, all ye apes and toads and worms, sure we are brothers — or, if not brothers, *parents and children!* The death that completely dissolves all the structures and compounds of our bodies must carry extinction to all our souls. We not only have no forever, but we have no future. We are only a better sort of cattle. All these inferences from the Law Scheme are of the million-to-one class.

But there is still another closely related inference which appeals formally to the Theory of Probabilities. If the Law Scheme is sound we ought to find neither men nor any other class of beings confined to any one standard of *intelligence:* but while men should show far higher specimens of mind than they do now, such high specimens should be quite as frequent among the other animals, and even among plants and lifeless things, as among men.

Do mental faculties come of a certain combination of certain blind atoms? Then, of course, these atoms on whose special properties and laws our minds ultimately depend, are, in the nature of things, just as open to one measure of these special properties and laws as to another. Why not? What is there in the nature of things requiring these elements to look toward the average human mind rather than toward that of a Newton or of a thousand-fold Newton? Why cannot they as well, from all eternity, beckon and bow in the direction of a Plato as of an ordinary mind, of a Jupiter as of a Plato, of God as of a Jupiter — why not propose to themselves One who can flood all space with sovereign knowledge and power as they themselves are already flooding it with their sovereign gravitation? I know of no man who is philosopher enough to tell why. But I know many men who are philosophers enough to see why they cannot tell. No reason can be given because none exists. Nor does any reason exist why such atoms, of whatever grade, together with such combinations as would make the most of them, could not just as well be connected with one organization as with another, with inorganic things as with organic. For it is not the *general* animal body that thinks and feels and chooses; this is structurally perfect even after it is quite dead. If the thinking power depends on a certain system of deftly arranged atoms, it must be a subtle interior system that eludes our nicest observation. And I say, there is no assignable reason why this occult system could not just as well be *connected* with one thing as with another, with inorganic things as with organic. Evidently not — as evidently as that there is nothing in the nature of a ship why it should move toward one point of the compass rather than toward another. The system of atoms, freighted with seeds and types and prophecies of the Academy or of Olympus, could sail away as easily and prosperously toward, and anchor as firmly by, a mollusk or a mountain as a man. All grades of these soul-germs, all conceivable combinations of them, all allocations of such combinations, are equally possible and easy to the thought — equally possible and easy to such a scheme of Nature as the Law Hypothesis supposes. As candidates for the world of self-consistent and scientific ideas, as candidates for the world of outward realities, there is nothing to choose between them. One would leap into Fact just as readily as the other at the bidding of a Creator. It is a case of unmitigated lottery. In advance of a given mind, it is a pure toss-up what measure of endowments it will have, and with what natural objects it will stand connected — whether it will have the twilight intelligence of the humblest brute, or the midday intelligence of the loftiest Christian archangel; whether it will be linked to a human body, or to the body of a tree or a body of water — as pure a toss-up as when a penny is shot aloft with the abandon of a Bacchus. Hence, in the case of a vast number of minds, we ought to find as many examples of one grade as of another; and we ought to find all the grades sown with impartial broadcast among all sorts of things.

Instead of each sort of animals having its own standard of intelligence, each should have the sweep of all the standards — instead of plants and inorganic things being totally without mind, they should be as liberally supplied with all its grades as are the animal tribes. Millions to one it would be so.

Now see the actual world. Is not every man conscious that he is a free, moral, and justly responsible being: and have not all mankind, without exception and from time immemorial, treated each other accordingly — praising, blaming, and punishing to any extent? — Are we prepared, despite the almost universal doctrine and hope of mankind, to give up the idea of Another Life — not merely the proof of it, but the very *possibility* of it — to give up all idea of ever meeting again our dissolved fathers and mothers, or even our own dissolved selves (to be dissolved a few years hence)? — Is there no standard human mind? Are men with the souls of foxes and geese and worms on the one hand, and of the traditionary angels on the other, as common as men with such souls as yours and mine? There is as much a standard stature for the minds of men as there is for their bodies. We find occasional dwarfs, occasional giants also; but the vast majority of minds are clustered closely about a certain mean measure toward which, it is plain, they all gravitate, and from which all deviations are due to disturbing forces. The oscillations of the pendulum recognize a fixed center. Just as the needle recognizes the Pole, though it has some play to either side of it; just as all the particles of a sphere recognize the common center of gravity, though some succeed in approaching it more nearly than others: so our souls, though variant within certain limits, recognize and incline to a certain common and fixed type of capacity. Where are the human Jupiters? Where are genuine Minervas, hiding in the flesh and blood of genuine Mentors? The pool, the pond, the Pacific — is our human tonnage as often gauged to one as to another? — So of the other animal tribes. Each has its own standard measure of the intelligent principle, and that a very small one. Any Houyhnhnms ever discovered by errant Gullivers? Any sign of Pitts and Websters among the oxen? Any hint of Tassos and Dantes among the fowls? Worms and insects — do they ever give token of being Napoleons and Massillons? When we speak of the brute creation, is it all an egregious misnomer; and, while the intelligence of one animal shakes the wing of a butterfly, does his fellow of the same species mount on that of an eagle, and still another fellow soar with a wing that can shadow a solar system? Do plants and inorganic things love and hate and plan and choose as well as the best of us? Do Mozarts and Ciceros and Sapphos wave to the breeze under the name of myrtles and oaks, or ripple over the stones under the name of brooks and rivers, or draw the clouds and lightnings under the name of hills and mountains, or trip our feet under the name of stumbling-blocks? Is the power of articulate speech all that is needed by the "things," to enable them to turn the Fables of Æsop and Fontaine into history — so that the bramble shall advise the cedar of Lebanon like a very Nestor, and storks and foxes and lions reason together as if the Seven Wise Men of Greece were in them? Is the old mythology, after all, nothing but a sort of Natural History; and is almost any grove, or island, or stream, or valley, or hill, possessed by an Intelligence which may be as worthy of love or fear as any sylvan deity ever fancied by the heathen? Are there Nereids for the seas, Naiads for the streams, Dryads for the woods? Is there a Ceres in the corn, a Pomona in the apple, a Bacchus in the grape, a Vulcan in the volcano, a Diana in the moon, an Apollo in the sun, and a Venus in the sweet evening star? Is everything really *possessed?* The metaphors and personifications of orators and poets — are these merest nineteenth century prose? Are oaks really stubborn, roses proud, lilies of the valley humble, vines affectionate, cypresses melancholy? Is there really a grain of sense in the child getting angry at and beating the stone against which he has stubbed his foot? Do landscapes really smile, the clouds get angry, fires and waters rage, stones and stars and all between understand the apostrophes we address to them? Ho, Father Tiber! Ho, Mother Earth! Ho, crescent Astarte! "Ho, thou that rollest above, round as the shield of my fathers — whence are thy beams, O Sun, thine everlasting light!" And is there really nothing to hinder us from getting satisfactory answer from these personages, save perchance some such slight difficulty as that of their hearing or talking our vernacular Latin, or Saxon, or Phenician, or Celtic? Has the shadow really gone back so far on the dial of old Chronos, and are we once more in the midst of the old heathenism. God forbid — if there be a God!

And yet this is what we should find in case the Law Hypothesis were true. Millions to one we should find it — pronounces the great Doctrine of Chances. She pronounces it from above that *Arc de Triomphe* which our best nations and sciences have united in raising to her honor, and covering with the marble pictures of her achievements. And yet, so far from finding minds of all orders, from that as small as the tiniest dew-drop to that as large as the hugest world, scattered at random among all sorts of natural objects as if from the hands of some Quadrifrons Janus, we find most classes of these objects showing no minds at all, and those which do show them confined each to a single standard capacity. Behold the measureless improbability with which the atheistic Law Hypothesis is weighted! Such a millstone ought to drag it down to the very bottom of the seas — there to lie drowned and buried forever. And this is the same thing as saying that it ought to raise that shining alternative, Theism, which, beautiful as an angel, presses foot on the other arm of the lever, as high as Heaven — thence to behold and rule all nations. Let us say, All hail, loftiest Doctrine! Infinite probabilities with an infinite exponent are swelling beneath thy buoyant and star-fanning wings! How keenly flash those stars under that mighty pulse! And then, full surely, no such discriminations ever came out of a dice box, or were reeled off from some great lottery wheel of blind law. I am sorry for you if you cannot see it — but lo, GOD! See here the true First of Aries! See here the true Prime Meridian of science! See here the true Epoch of history for the whole astronomical heavens! It was doubtless a true Natural Selection that presided over the grades and distributions of Mind; and gave to man his day, to the worm its dawn, and to innumerable objects most fair and exquisite in their way their starless night — a true Natural Selection — but then the name is too long and savors too much of a mere *Thing*. Please call it GOD! And GOD it is — personal, eternal, unbounded — though some few men may choose to hide and belittle and suppress Him under the learned names of Pangenesis, Parthenogenesis, and Protoplasm.

VII.

CONFLICT WITH SOLAR ASTRONOMY.

Nam cum dispositi quæsissem fœdera mundi,
Præscriptosque mari fines, annisque meatus,
Et lucis noctisque vires : tunc omnia rebar
Consilio firmata Dei. — *Claudian.*

VII. Conflict with Solar Astronomy.

1. NEBULAR HYPOTHESIS 197
2. ESTIMATED 199
3. CENTRAL HEAT 203
4. CHEMICAL CONSTITUTION 205
5. MECHANICAL RELATIONS 209
6. ROTATIONS 216
7. REVOLUTIONS 223
8. WHAT NEXT? 239

SEVENTH LECTURE.

CONFLICT WITH SOLAR ASTRONOMY.

IT is essential to the Law Hypothesis that it account, not only for the living organic bodies on our globe, but also for that globe itself and all those systems of worlds which make up the great realm of Astronomy. This it attempts to do.

Thus. Suppose out in free space a great mass of glowing vapor. This vapor, under the influence of gravity, will take a globular figure, as being the figure of equilibrium. This vaporous globe, by loss of heat and stress of gravity, will gradually contract on its center. This inward flow of matter from all quarters will be sure to be made unequal here and there by inequalities of structure, and so will cause a central whirlpool. This central whirl will gradually extend itself to the whole mass and become its rotation. This rotation will cause an equatorial protuberance. This protuberance will gather speed as the mass contracts until, at last, the centrifugal force will exceed the centripetal and a ring be detached; and, as the contraction goes on, successive rings with an ever increasing speed of rotation. Each of these rings will have in it some nuclei — one or more — of special condensation and confluence of currents, and so eddies: and these nuclei will finally draw the whole material of the ring about themselves and become so many rotating planets; each of which in its turn may throw off from its equator successive rings to become satellites. So, at last, a solar system may be born out of mere vapor by the sole means of well-known natural forces and laws.

This is the noted Nebular Hypothesis. Its more advanced friends add that the system thus naturally constructed must at last be ruined in a natural way, if not by the gradual decay of the solar heat, by the resistance of the ether which is found occupying the planetary spaces. This will retard the planets and satellites in their revolutions and at last cause them all to fall into the sun, there to be vaporized and expanded again into the original fire mist. Then, in the same way as before, the system will be built up anew. And so on eternally. And so backward eternally.

See a specimen of the history of every celestial system! The nebulous tree slowly ripens its fruit through the ages until comes the cosmical autumn. Then Law lifts its great hand and shakes the loaded boughs till they have cast all their mellow and yellow globes — as it is written, The stars shall fall from heaven and the powers of heaven shall be shaken. But at length comes the cosmical spring again; and new world-buds, blossoms, fruits take the place of the old and exactly repeat their history. And so on, in endless repetition, upon the same huge immortal tree.

In support of these views it is claimed that fire mists are actually found in space — that they are found, apparently, in various stages of such a world-forming process as has just been described; condensing at the center, parting into nuclei, acquiring rings — and that, on the supposition that the solar and stellar systems were once fire mists and formed according to the Nebular Hypothesis, their leading facts are well accounted for.

Before considering how far the hypothesis and its alleged proofs agree with facts, it is worth while to notice that the agreement might be perfect, and yet amount to very little as support to a scheme of mere naturalism. Most theists are free to admit that there are such things as natural causes. They are also free to admit that such causes are able to do not a few interesting and striking things. If any one can show that among these interesting and striking things is the compacting and arranging of those huge masses which we call planets, suns, stars — what follows ? That mere Nature can do everything ? By no means. That mere Nature can people these worlds which it has made with the higher wonders of vegetable and animal and spiritual life ? By no means. All that follows is that I must enlarge somewhat the field of possible natural causation, as it lay in my mind. I have been inclined, say, to regard that field as bounded by terrestrial crystals ; now I must regard it as including those huger crystals in space which we call worlds. That is all. I have merely shifted boundaries a little — I have not abolished them. I have advanced the walls somewhat, but I have not advanced them beyond the entire universe of matter and mind. A great domain is still left for the supernatural. Still may it be true that this domain includes all that living world of organic and spiritual being whose intricate and exquisite adjustments and powers, even in their lowest forms, baffle our understandings as no astronomic systems have ever done.

Proving the Nebular Hypothesis would amount to very little. But disproving it would amount to a great deal. It would break down the whole Law Scheme, break it down with a feather, break it down at a point where it really is strongest. Nowhere does that scheme have so specious a look, nowhere does it really come so near to adequacy, as where it attempts to account for the origin of worlds. And worlds of some sort are by far the easiest part of Nature to account for on natural principles. No one ever yet had a glimpse, or pretended to have, of the way in which atoms could naturally come together so as to make a Newton, or even a daisy ; but almost any one can see how atoms might come together, through gravity and loss of heat, into a great round body of rude matter for a daisy and Newton to live on. It can be made clear to a little child. Still it is a question whether the actual worlds (which are something more than great heaps, or even happy compilations, of moving matter) were *actually* made in this way. If it can be shown that they were not, then the Law Scheme is shown unequal to the easiest of all the work it has to do.

Can it be shown — shown that such an origin

of the worlds as the Nebular Hypothesis supposes is in direct conflict with astronomical facts?

That it does not conflict with *some* of these facts is a plain matter. No doubt there are many bodies in space more or less aeriform; that many of these bodies are incandescent; that some of them have shapes and aspects such as they would have if ripened from a vague fire mist after the manner supposed; that aeriform incandescent bodies of a certain sort would, in virtue of well-known laws of heat and gravity, gradually condense into solid spheres having revolutions and rotations somewhat like those of our own system; and that, at last, such a system might come to a fiery end in which "the heavens dissolve and the elements melt with fervent heat." All these, and some lesser facts to the same effect — such as the agreements of the Nebular Scheme with the particular shape of our earth; with the gradual increase of heat toward its center; with the traces of a far higher ancient temperature at the surface than now exists, and even of a general igneous fusion — are freely granted. There is, beyond a doubt, a large and interesting agreement between the heavens as they are and the heavens as they would be if built according to the Nebular Hypothesis.

But, then, this is only one side of the subject. There is also another large side — a side of disagreement and insufficiency. Not only does the hypothesis fail to account for many astronomical facts; but it is in direct conflict with many other astronomical facts of the surest and weightiest character. I will give some examples. And they will be, on the present occasion, exclusively from the Solar System.

I begin at the center of the System; and call your attention to the *immense heat and brightness of the Sun.*

If the Sun is the residuum of a fire mist which has been cooling for an immense period, its original heat and brightness must have been almost infinitely greater than they are now. But what *is* a brightness almost infinitely greater than now belongs to the Sun — on the face of which the Voltaic arch, whose as yet unmeasured heat vaporizes the most refractory substances, turns to a shadow? Something inconceivable. Something wonderfully beyond example — unless the nebulæ give examples. But they do not. The latest advices from the spectroscope — as I shall have occasion to notice more fully hereafter — are to the effect that the gaseous nebulæ, so called, are not particularly heated. Besides, if they were, they would, as a rule, outshine those nebulæ pronounced solid by their spectra; as they do not. Moreover, their central part would outshine the stars themselves; on account both of its greater size and of its greater intrinsic brightness. We ought to see at the heart of every large gaseous nebula a region of visible diameter so effulgent as to put to shame most easily the glory of Sirius himself. It should be so — unless the nebulæ are in general much further from us than are the stars. But this would not consist with the view that the stars are all made from fire mists, and are destined in due course of Nature to turn to fire mists again. This view requires that the gaseous nebulæ, on the average, be at the same distance from us as the stars which stand for their completed state.

Thus the Solar System, at its fervid center, begins to object to that hypothesis which, whatever may be thought of its real significance, is the sole dependence in these days of all who try to explain the natural without help of the supernatural. Instead of falling down before an atom (say *several* atoms, if you please) and saying, Thou art my Maker, the Sun prefers to go further for its worship and fare better; and sends far down the abysses its flaming glances asking, Where is *He?*

I next call your attention to the fact that our Solar System has not *the same chemical constitution* throughout — as it would have if all its members came from the same fire mist.

By the laws of gravity the materials of the mist would tend to arrange themselves in the nebulous sphere in successive layers, according to density, the densest at the center ; but, by the laws of heat, the heat in such a sphere would become supreme at the center where its loss is least, and so start thence to the outside currents that, coming and going everywhere — like the streams of pilgrimage at Mecca, center of faith ; or like the streams of trade at Palmyra, center of gain, in her days of palm — would carry everywhere on their incessant caravans all the elements of the nebula and intimately mix its constituent materials. Of course, other heat-centers would still further promote the mixing. It would be unbounded free trade. Every place would get everything. It would be a case of extreme boiling — of the most thorough, most terrible, and most merciless sort. As every housekeeper knows, the mighty caldron could not fail to get its various contents well distributed. In no other way can such chemical unions as we find throughout the Earth, between the lightest and the densest substances, be accounted for by the hypothesis. In no other way can it account for the present state of the Sun, whose tempestuous atmosphere of hydrogen, the lightest substance known, is found by that greatest of detectives, the spectroscope, to be charged with the heaviest of known metals. Considering the Sun as the residuum of our fire mist, it could not be so flooded with hydrogen as it is, unless that element had pervaded the whole original nebula. It would all have been used up in manufacturing the planets. Indeed, hydrogen has been found picturing itself most strongly in the spectrum of Uranus.

So the great nebulous egg, out of which by the incubation of heat and gravity the broods of our Solar System are said to have been hatched, must, like other eggs, in due course of incubation have had its contents thoroughly stirred and churned and mingled. It must have come to the same general constitution throughout. And, becoming organized into worlds under quite the same general conditions throughout, it must have yielded everywhere bodies of the same general character as to kind and proportion of materials.

Are the members of our System actually such bodies? We go to the spectroscope for answer. This subtle analyst confesses that terrestrial elements are very largely diffused through the solar realm, but it also tells us of great differences in the chemistry of its various bodies. For example, it tells us that oxygen and nitrogen, both of which so abound in the Earth, and one of which makes up not far from half its substance, either do not exist at all in the Sun or do not exist in any appreciable proportion — while chromium and other metals, which make no figure in the composition of the Earth, show themselves in great force in the solar atmosphere. Some fourteen familiar substances write out their names in a plain hand in the spectral register: but most of the names we read there are the names of perfect strangers. Of the two thousand lines that cross the solar spectrum, by far the greater part have nothing whatever answering to them among terrestrial elements.

The comets, also, as far as examined, show a chemical treasury quite different from all other bodies of our System. Of the four comets whose spectra have been criticised, two consist of unknown elements ; while the other two consist, one of carbon and the other of nitrogen. Here we have a triple unlikeness to the Earth and Sun -- but a single element belonging to each comet, that element different for each, and in half the cases examined that element quite new to our chemistry. How could a single element like carbon disengage itself from the confused stirabout of the parent mist, and set up a sphere and revolution of its own — especially *such* a sphere and revolution? The idea that comets are foreign bodies will be considered further on.

In regard to the planets — it is found that the solar light as reflected from Mars, Jupiter, Saturn, and Uranus gives as many different spectra, each differing somewhat from the direct solar spectrum ; showing that at least the atmospheres of these planets are to some extent differently constituted. And a difference in the atmospheres implies a difference in the bodies beneath. We sometimes say that a man is known by the company he keeps. We are quite sure that we can divine his inner character, if not from the way he dresses, at least from the kind of atmosphere he breeds about himself. So we can divine in the case of the planets. Different atmospheres betoken different interior chemistries. In this connection the case of Uranus is particularly striking. Its envelope consists largely of hydrogen, but is quite without sign of oxygen. Were oxygen there in a free state, a single spark would set the planet aflame.

So the Solar System objects again to that hypothesis which, whatever may be thought of its real bearing, is in fact the last stand and dependence of all who in these days try to explain the natural without help of the supernatural. This time the challenge comes from the *Chemistries* of the System. Instead of falling down before an atom (say a fog-bank of atoms, if you please) and saying, Thou art my Maker, these Chemistries prefer to go further for their worship and fare better; and send far abroad into space their messenger-spectra, robed like Solomon, asking, Where is He? For *that* He is, they can have no manner of doubt. The Law Scheme breaking down, there must be God. As science now stands, there is no *tertium quid* — God is the only alternative to development.

Consider *certain mechanical relations.*

By these I mean the relations of the principal

bodies of the Solar System to each other in regard to such matters as size, density, atmosphere, position, number of satellites.

One well-mixed material forming the fire mist. This one material acted on throughout by two causes, gravity and heat ; the one steadily increasing in force at the outside, and the other as steadily diminishing. This one material steadily becoming smaller and denser, steadily increasing its speed of rotation, steadily throwing off rings that steadily become less. So runs the hypothesis. That is to say, it supposes that the members of the Solar System are formed successively out of the same material, by the same causes, under the same circumstances, or such as vary steadily in one direction according to a simple law. Of course the products should be alike ; or, in chief points of unlikeness should vary in the same steady, straightforward manner as do the circumstances on which they depend. Thus, if one member of the System is without atmosphere and water, all the other members should also be without them. If they differ in density, the series ought to show a steady increase of density toward the Sun. If they differ in size or number of satellites, the series ought to show a steady decrease toward the Sun in these respects. So it would seem.

What are the facts? We turn to our Moon and find it without sign of water or atmosphere; while the Earth from whose surface it was cast off is well supplied with both. We turn to Mars and find that, though god of war, he has not a single henchman to attend him, not a single page even; while his two nearest neighbors worth mentioning, on either hand, are provided for; indeed, all the chiefs of the System beyond himself, nobly, though irregularly, provided for. Running eye along the whole bright line of orbs, we notice that in general those near the Sun are smaller and denser than those more remote; but, just as soon as we come to particulars, we find we must break up completely the actual order of succession that we may range the planets in the order of their size or density. Mars is smaller than the Earth or Venus, though further from the Sun. Saturn is smaller than Jupiter, and yet larger than either Uranus or Neptune. The great Giant of the System, as if to intensify his stature, takes stand by its veriest dwarf. So of the densities. There are Uranus and Neptune both denser than Saturn. Here is Venus no denser than Mars and not so dense as the Earth.

So of the orbital eccentricities and inclinations.

So of the rates of rotation. As we follow the planets down to the Sun, in the supposed order of their age, it is all backward and forward after a most capricious fashion — this is the way we speak — with nearly all the chief mechanical elements of the System. The same perplexing irregularities exist in the satellite systems, if we may take the Jovian family for an example. In place of that steady variation in one direction which answers to the hypothesis, we have one strongly like, in its break-ups and reversions, to the apparent path of a planet. In short, as Humboldt says, "The planetary system, in the mechanical relations between its members, does not appear to offer to our apprehension any stronger evidence of a natural necessity than does the proportion observed in the distribution of land and water on the Earth, the configuration of continents, or the height of mountain chains."

As to the position of the solar bodies among themselves — they ought to lie in successive, well separated *groups*, answering to the successive rings of vapor cast off from the nebula. Thus, there should be several planets in the general district occupied by Neptune; after an interval, another similar group in the general district occupied by Uranus; and so on. The solitary should be set in families. The planets should show themselves of a social turn. And the satellites, in this respect, should take after their parents. As humanity divided itself and went forth into the early world by clusters of related families; as northern Europe in our day breaks up and appears again by societies of kin and neighbors in our western wilderness — so, on nebular principles, the cosmical vapor should have parted and gone forth to the peopling of the void by *clusters*, rather than by individuals. For, it is not to be supposed that a ring would generally have but one nucleus; or that, there being several nuclei, one of them would generally prevail over and finally absorb all others. The chances are greatly against either event in any given ring. As should be admitted with special readiness by those who admit that some comets, and the great meteoric systems, and the rings of Saturn, are, or may be, composed of myriads of distinct bodies.

Consequently, there ought to be in the System several cases, at least, of a ring being resolved into two or more worlds: whereas there is only one case that can possibly be supposed to be such, namely, that of the asteroids. The chief planets are all apart by immense intervals, such as could not separate bodies born of the same ring. Each planet is a hermit. Each satellite, also, so far as known, is a hermit—a very Simeon Stylite, separated both by horizontal wastes and wastes perpendicular from all his fellows. And as to the clan of asteroids—if we admit that they all came from one ring, in spite of the great difference between them in orbital planes, then it follows that a ring may part into more than one hundred bodies of unequal size and contiguous paths and paths wonderfully interwoven, which yet can keep apart indefinitely long: hence, that it is very unlikely that in all other cases, not to say in any other case, a ring has furnished but one world, or one world has devoured all the rest. The parent nebula may be expected to colonize by companies, not by individuals. Like the wicked man, it will send forth its children by *flocks*. It will scatter itself in successive showers, instead of a continuous drizzle. The planets will be Jews: eight tribes, if not twelve, from center to outskirt will pitch nomadic tents. Gregarious planets, gregarious satellites—each class found, for the most part, in distinct groups, instead of each orb appearing as the solitary heir and representative of its own ring. Such should be the "manner of the kingdom" — the kingdom over which blazes the corona of the Sun, of which the planets are the great Peers, and of which millions on millions of comets and meteors are the multitudinous Commons. To suppose anything else, would be about as unreasonable as it would be for a statistician to suppose that Irish families, on the average, turn out but one child apiece.

So the Solar System continues to object to that hypothesis which, whether it mean atheism or not, is undeniably the last stand and sole dependence and loud boast of all who try to explain the natural without help of the supernatural. This time the challenge is many-voiced and comes from the *Mechanical Relations* of our System. Instead of falling down before an atom (say, if you please, a great *cloud* of atoms) and saying, Thou art my Maker, these great Relations prefer to go further for their worship and fare better, and hang out signals to the infinite Beyond from every promontory and hill-top and turret of the System (whether its name be density, or size, or atmosphere, or satellite) asking, Where is He?

For *that* He is, they can have no manner of doubt. Where the Law Scheme breaks down there must be God. There is no *tertium quid.* As science now stands, a personal Creator is the only alternative to eternal evolution. Who to-day believes in chance? A man, a Solar System, a Paradise Lost, made by a fortuitous concourse of atoms — who now believes that, or is in danger of believing it? Not a solitary soul. It is all law, eternal law, with present unbelievers. And when law breaks down there is nothing left on which the reasonable thought can fall back, save "the King eternal, immortal, invisible; the only wise God."

Consider, too, the *rotations of the Solar System.* We find that these rotations are not all accurately in the planes of their respective orbits; nor is the law of their periods the same for both planets and satellites; nor is the parallelism of their axes to themselves, respectively, disturbed in the course of a revolution — all of which facts, each including many particulars, attack the Nebular Hypothesis. And it is a matter worth pointing out that, in this attack, the assailants march by bands, and not by individuals. It is not so many separate soldiers that we see — it is the Macedonian phalanx. It is not a war of guerrillas that we have — it is rather a disciplined concert and chorus of battles. The clans have gathered, the sections are moving, it is almost "the uprising of a great People."

Under this head, I ask you first to notice *the ring of Saturn*. We are told that in this ring we have an actual example of those nebulous zones which play so great a part in the nebular cosmogony — an *actual* example which Nature has kindly stereotyped and preserved as an illustration of her way of building worlds. If this is so, I answer, the exterior division of this ring, whose distance from the planet is nearly fifty thousand miles, ought to rotate considerably slower than the planet; in accordance with the law which requires that the radii describe equal areas in equal times. Actually, however, it completes a rotation in about the same time as its primary.

Now go on to notice that the rotations of our System differ in the *principle of their periods* for the planets and for the satellites. These periods are smaller for the more distant planets and larger for the more distant satellites. And, so far as the satellites have been well studied on the matter, they all seem to rotate in about the same time as they revolve ; while no such law is found binding the planets. Thus, our moon turns round once on its axis while it is making one journey about the Earth. So also the moons of Jupiter seem to do ; keeping always the same side turned toward their primary. But not a single planet is known which makes a circuit on its axis in the same time it makes a circuit round the Sun. A very singular difference ! How could this be if the System was formed throughout in the one way taught in the Nebular Hypothesis ? For, according to that hypothesis, the satellites come from their primaries in precisely the same way and under precisely the same general conditions as these primaries do from the Sun. Hence the same general law which governs the rotation-periods of the one class of bodies should govern those of the other class.

Add to this that the rotations of the System are not all *exactly in the planes of their respective orbits*. Far from it. The Earth rotates at an angle of about 23°, Saturn of 26°, Mars of 28°, Venus of 50°, Neptune of 76°, and Uranus of 100° ; while not a single planet or satellite is known to rotate exactly in its own orbital plane. All seem to have made a solemn League and Covenant with each other *not* to do it ; and keep to that Covenant with even more than Scottish determination.

How can this agree with the hypothesis ? If the System had been formed in the way supposed, all these angles of rotation would have been zero. For, suppose a rotating gaseous ring, and a nucleus entangled in it. A body so entangled tends to have one rotation in the plane of the ring in the course of one revolution — let our moon stand for an example — and, by the gradual condensation of the ring, tends to have this rotation hastened. The gradual shrinkage from all quarters on the body, besides pressing it toward the middle of the ring, will, according to a well-known law, hasten its outer surface and retard its inner in a direction parallel with the middle plane of the ring, and so tend to make a rotation in that direction. This threefold force, being both strong and persistent through vast periods, and the last to act on the planet as it becomes detached and solidified, will at last wear out any rotation in a different direction that has chanced to fasten on the nucleus. For such a rotation depends on a force necessarily brief and changing ; and is steadily resisted by the whole neighborhood of atoms relatively at rest, and by the general equatorial drift and suffrage of the mist; and so can never amount to much, or be permanent.

So strong a partnership against one so weak cannot fail at last to have everything its own way. It will by degrees gather to itself all the business. Its rivals will disappear. Even that general equatorial drift alone, a plenty of time being allowed it, would suffice to suppress all other rotation in favor of one in the same plane and direction with itself. Then, in the last stages of the nucleus, as it becomes a discrete and solid orb and issues its Declaration of Independence, comes in the zone-condensation with its powerful forces to confirm the result. As is very common, just as soon as the victory is gained help comes. Who of us is not free and proud to help the man who no longer needs help? At last his cup runs over. Assurance is made doubly sure. So here. And all westerly rotations — which, apart from the equatorial drift, are just as likely to occur as the easterly — are stopped and reversed; and all rotations out of the plane of the orbit are reduced into it; and the whole orb is put under heavy bonds for the future to remain in the state it has reached. So it must be on the principles of the hypothesis: and so it is very far from being in the actual System.

I have stated that the *parallelism of the rotation to itself* is not disturbed during a single revolution. So far as astronomers have been able to notice, this is a fact; and it is universally taken for true of all the members of the System. But how could it be true if the System was formed in the way supposed? In all cases where the axis of rotation is not perpendicular to the orbit, there must be a change in its inclination to a fixed plane, during a revolution, proportional to the smallness of the inclination. Where the inclination is nothing, the axis will completely reverse itself in the course of a revolution — as the spoke in a wheel does, as it revolves. Suppose a nucleus entangled in a revolving gaseous zone. Let its axis coincide with a radius of that zone. Then the nucleus will be borne round with the interior end of the axis pointing always at the zonal center — as exactly and steadily as do the lines of gravity; or, what is the same thing, as exactly and steadily as do all the great Christian doctrines at the Cross — so that, at opposite points of the orbit, the same end of the axis will point in directly opposite directions. If the axis, instead of being coincident with a radius of the zone, is perpendicular to it, then there will be no change in its direction during the revolution: it will always remain parallel with itself: and, in case all the revolutions of the System are in the same plane, all the swiftly going solar troops will carry their ghostly spears in exacter parallelism with each other than was ever insisted on by the martinets of our earthly armies. With the axis anywhere between perpendicularity and coincidence, there will be a change in direction inversely proportional to its inclination. But, as in fact the axis is generally far out of the perpendicular — like human nature itself — it must generally undergo a great change of inclination; and the whole solar cavalry will carry their weapons after as disorderly a fashion as ever did rawest recruits from the rawest provinces.

So the Solar System does not yet weary of objecting to that hypothesis, which, whether it denies a God or not, is undeniably the last stand and whole dependence and loud boast and best helper of all who now try to explain the natural without help of the supernatural. This time also the challenge is many-voiced, coming as it does from the many *Rotations* of our System. Instead of falling down before an atom (I am quite willing to say a *nebula* of atoms, if you choose) and saying, Thou art my Maker, these Rotations choose to go further for their worship and fare better; and point every axis with the pointed inquiry, addressed to all the heavens, Where is He? For *that* He is, they can have no manner of doubt. Where the Law Scheme fails there God must be supplied. There is no *tertium quid.* As science now stands, an eternal Creator is the only alternative to eternal evolution. Who to-day believes in chance? A Newton, an astral system, a Mecanique Céleste made by a fortuitous concourse of atoms — who now believes that, or is in any danger of believing it? Not a solitary soul. It is all law, eternal law, with the latest unbelievers. And when law breaks down, there is nothing on which the reasonable thought can fall back, save "GOD, the Lord, He that created the heavens and stretched them out, He that spread forth the earth and that which cometh out of it, He that giveth breath to the people upon it, and spirit to them that walk therein."

Finally, consider the *revolutions of our System.*

These are not all in the same direction, nor exactly circular, nor exactly in the plane of the Sun's equator — as they ought to be if the System was formed after the manner of the Nebular Hypothesis. On the contrary, none of the bodies composing the System fully meet all these conditions; while many of them are very far removed from doing it, and some may even be said to *trample* upon them as they go their shining way.

Thus, there is not a single body known to have a strictly circular orbit. Some orbits come very near being circles, but all are really ellipses. In some cases the ellipses are very eccentric. The eccentricity for Mercury is about one fifth of the semi-major axis of its orbit, that for Mars one tenth, that for Jupiter or Saturn or Uranus one twentieth. One of the asteroids, Polyhymnia, has an orbit even more drawn out than that of Mercury. But the comets and meteoric systems are the most surprising objects. Several of them move in paths many times longer than they are broad, and some sweep about the sun on curves almost parabolic in character. Encke's comet is twelve times nearer the sun at its nearest point than at its most remote. Halley's is fifty-seven times nearer; and some of these wonderfully eccentric bodies almost brush the Sun with their fiery tresses, and then, as if in mortal terror, fly away far beyond the orbit of Neptune. The August and November meteoric systems almost rival the most eccentric comets in their paths ; both approaching the Sun as near as the Earth, and then going away from him, the one further than Neptune and the other than Uranus. How can such feats of cosmical knight-errantry agree with a scheme that supposes all the members of the System formed out of exactly circular rings, all of whose matter moves in exact circles, and which were detached from the solar nebula at the point where the centrifugal and centripetal forces were equal? Under these forces in equilibrium, a body must describe a purely circular orbit, instead of one of those mighty ovals which almost promises never to return into itself.

So *they* virtually admit who appeal in support of the Nebular Hypothesis to the generally "nearly" circular form of the planetary orbits. "See," say they, "how little the path of Venus differs from a circle: this is as it should be if the worlds had a nebular origin as supposed." Yes, we see it — but what of the many great exceptions? Yes, we see it — but why say "nearly" circular? The orbits ought to be *quite* circular — every one of them ought to be quite circular — to meet the demands of the Nebular Theory: and instead of the comets running such amazing tilts as they do into Nox and Erebus, as if at the fixed stars, they ought to go about the Sun on paths as round and at distances as unchanging as if they were fastened to the solar center, each by such an inflexible golden chain as Milton gives to the Earth.

Besides, according to the hypothesis, the revolutions of the Solar System ought, without exception, to be *exactly in the plane of the Sun's equator.* The friends of the hypothesis should admit this easily. It is almost one of their own doctrines. They say that the solar orbits ought to show a general partiality for the solar equatorial plane: and they loudly appeal to the fact that such a partiality exists, in support of their views. "See," say they, "how closely Jupiter and other planets keep to that celestial loadstone, the ecliptic: this is just what we should find if all the bodies were turned off from the solar equator in the manner supposed." Yes, we should find that — and just a little more. We should find not only a general partiality but a *universal* one. We should find not merely a partiality, but such a partiality as expresses itself in absolute union. Coquetting between the orbits is not enough. Hard wooing will not answer. Marriage must be consummated. And such a marriage as never asks for a divorce, or at least never gets it, not even from the State of Connecticut. All the orbits, without exception, ought to lie, exactly and forever, in that one plane in which the sun rotates, and along which its expulsive power always acts to cast off its children, when their time for leaving home and setting up for themselves has come.

Let us inquire a little into the early *history* of these celestial outcasts. It is supposed that all the nuclei which happen to form in any part of the parent mist are gradually drawn into its equatorial zone before they ripen into solid worlds. It is necessary to suppose this. A group of revolving worlds cannot be formed by the breaking up of a mist into several sporadic nuclei in advance of a general rotation : for, such a rotation must begin at once on the formation of the mist from whatever cause ; also, without it, such nuclei being stationary in respect to each other and the center of gravity of the mist, would gradually settle in straight lines on that center — it being necessary to revolution about that center, on the part of any body, that it should have a motion across the straight line joining the two. Nor can such a sporadic group be formed after general rotation begins. For, conceive the mist with its rotation and bulging equator well established. This bulge rises gradually from the poles and reaches its highest at the equator. When a part about this highest point is thrown off as a ring, the sides fall in on both hands and swell out into another equatorial bulge. So an immense suction is established. Permanent currents set in from both poles toward the equator. They are the Trade Winds of the mist, and by them whatever nuclei may be near the surface, away back to the poles, will be driven, like so many ice-bergs which they are not, or like so many fire-bergs which they are, toward the equator. Of course these are the nuclei that have the greatest ripeness, because they are the nearest the condensing cold space; and the ripest of all are those nearest the equator, and especially those in the detached equatorial ring, because these are most remote from the center and most exposed to the condensing cold. The great nebular tree, like other trees, has its most advanced fruit on the outside. Thus the nuclei near the surface, all over the mist, will be gradually drawn, in the order of ripeness, into the equatorial regions, as ring after ring is thrown off.

Of course the same influences that stress the cosmical embryos toward the equatorial zone, and secure that every one of them shall reach it before ripening into a solid world, will finally carry them all to the central plane of that zone, which is the final goal of all the influences. This is the plane of form equilibrium, which always tends to become the plane of the center of gravity. It is also the plane toward which condensation by cold tends to drive the atoms; also the plane of motional equilibrium; in fine, the plane toward which the whole System, by all causes of movement that can act on it, strains. It is the center of the wheel where all the spokes meet, and where they are all held fast by an iron band. This is admitted, virtually, by the friends of the Nebular Hypothesis when they appeal to the exceedingly thin ring of Saturn as an example of a ripened zone — a ring forty miles thick and one hundred thousand miles broad, and, in their view, composed of discrete bodies. Hence, in the immense course of ages spent in the ripening, each planet-seed will settle accurately into this equatorial plane, and there go on developing into the great world umbrageous with the mysteries of organism and life. Really, according to the view of atheistic friends of the Nebular Scheme, these tendencies toward the central plane have had an eternity to work in : for, these men have to suppose that all the past has been occupied in alternate constructions and destructions of the System under the constant pressure of these tendencies. Suppose the planets to gradually approach the Sun through the resistance of an ether pervading the System. Wheeling about their flaming goal in ever narrowing rounds and at ever increasing pace, the cosmical chariots at last strike against it, and, with broken axle and shattered frame, disappear from the racecourse. Up flames the solar bonfire fiercer than ever! Out swells the fiery nebula till the whole solar amphitheater, and more, is ablaze! King Arthur is dead, and each knight of the Round Table. But Arthur and his knights shall live again, that great Table shall be reset, and then each hero shall sit at it more squarely than ever. But what happens meanwhile? As the planets approach the Sun the inclinations of their orbits to the solar equator gradually lessen under the action of its protuberance : at last they fall into the Sun with a motion more nearly than now, if not exactly, in its equatorial plane, and with a velocity the same as that of the equator. This fierce fall itself must alter somewhat the plane of the Sun's rotation, and bring it nearer the orbit of the planet ; and, in the resulting fire mist, the planets must be represented by rotating nuclei more equatorially disposed than at present. So the approach to the equator would go on, through successive reconstructions, till at last the approach would become an arrival, and all the planets move accurately in the plane of the Sun's equator. At last the grand climacteric is reached. By littles and littles the planetary millennium has come. Henceforth shall nothing disturb the Concord of the Orbits.

In the same way it may be shown that the *satellites* ought to move accurately in the equatorial planes of their primaries. But these, as I have shown, ought in all cases to coincide with their orbital planes, and so with the plane of the Sun's equator and of the ecliptic.

So the hypothesis demands. But, on looking at the actual heavens, we find the demand not acceded to in a single instance ; and, in some cases, flatly and even sternly refused. Not a planet joins the Earth in moving exactly in the plane of the ecliptic. Pallas crosses it at an angle of 34°. The satellites of Uranus, and probably Uranus itself, cross it almost perpendicularly. As much is suspected of Neptune. As to the comets and meteoric systems — they have a sovereign contempt and perpendicular aversion for all rules in the matter; and, in their headlong steeple-chase through space, tear across our plane just as may happen to be convenient, whether at 9° of inclination or at 90°.

Next look at the *direction* of the revolutions in our System. This ought to be the same throughout: because all the revolutions are supposed to come from one cause, namely, the rotation of the solar nebula. This rotation must throw off the planets in its own direction (as who does not know who has seen the grindstone casting off its drops and sparks, or the millstone its grain): these, in their turn, must throw off satellites in the direction of the planetary rotations. But these rotations, as we have seen, should, according to the theory, be in the planes of their respective orbits, and so in the plane of the Sun's equator. But actually they are not. What we are entitled to find we do not find. While an easterly motion is the rule, there are many striking exceptions. The satellites of

Uranus move from east to west: probably the primary does the same. Very considerable reason exists for thinking that this westerly motion is shared by Neptune and its satellites. It certainly is by the meteoric system of November. And not a few comets join in this retrograde, and run backward on their orbits faster and more fiercely than ever planet ran forward — as backward going people are apt to do.

It is commonly allowed that many of the comets could never have come from the same fire mist as the planets. And no wonder. How can one look at those immensely elongated and almost upright orbits, as well as at the singular chemical constitution and fierce retrogrades of the ghostly bodies that traverse them, and do less? Not La Place. He said they must be foreign bodies. They must be importations from beyond our solar seas. One day our System in its progress through space neared these lost Children of the Mist, as they wandered about the wilderness spaces without a protector, and benevolently took them into its own family — despite their extravagant and undisciplined ways. Such Phaetons, such break-neck riders, such incorrigible vagrants, were enough to corrupt the habits of the staidest young family in Planetdom ; and yet the pitiful sire opened his arms and took them in. This was very kind of him — for aught I can see, impossibly kind. For, according to the Nebular Hypothesis these stray comets must have belonged at first to *some* fire mist and its system of worlds ; and, once members of such a system, how could they have been parted from it and united to our System, save by such a near approach of the two to each other as must have immensely changed the economy of both, and probably kept them together ever after ? Systems fairly brushing each other — especially sporadic ones with slow motions — would not part company in a hurry. They would not part at all. But there is no evidence that stellar systems have ever grazed each other in this manner — much less that it has been a frequent event, as it would need to be, considering the great number of comets. Certainly our System has never been party to a single such event within the scope of history and tradition. We are now, and, to the best of our knowledge, always have been, apart by almost immeasurable intervals from all the stars : indeed, it could not be otherwise consistently with the hypothesis which in the last resort derives all the systems of space from one great fire mist, and of course relates them to each other very much as the planets are related, only on a far vaster scale of intervals — while even the planets are so far apart that not one of them is able to draw off bodies belonging to another. They feel its influence, but are not overcome by it. Like the best men under temptation, they may round out their orbits a little toward the tempter; but yet they go firmly on their way, holding fast to the company and course to which they belong.

It should be noted that no *disturbing* influences of the planets and other members of our System, among themselves, can account for either their reverse motion in the orbits, or their eccentricities, or their variety of plane. Plainly not for the retrograde motion. Almost as plainly, not for their variety of plane: for, if all the bodies had originally been set revolving in the equatorial plane of the Sun, their actions on each other, being always within that plane, could never tend to draw from it. So with the eccentricities. The orbits, though flexible to a certain extent, are not like some hoop of rubber which one can draw out into a most eccentric oval without breaking. See how the boy strains between hand and hand! The circle has become almost a line! But the orbital diameters of our System can only vary from age to age by the merest trifles. La Grange has shown that the greatest possible change in the eccentricities, from the mutual actions of the System, is extremely small — in some cases less than the eccentricities themselves at their smallest. For example, the least eccentricity of Mercury is 0.1886, and that of Mars 0.0746. But the changes in these numbers for two hundred thousand years, reckoned from A. D. 1800 as middle point, are respectively only 0.0170 and 0.0347. In fact, the theory of gravity gives the least change to those small planets which have the most eccentric orbits. Surely it is not a remarkable merit in an hypothesis that it gives the least explanation where the most is needed! Besides, if the eccentricities were due to the disturbing actions of the worlds on each other, the fluctuation would be about a circle as the mean figure. But the mean figure is not a circle, but an ellipse. The eccentricity is never zero in a single solar orbit. By no means the only class of orbits of which this can be said!

So once more the Solar System shakes its

bright locks horizontally at that hypothesis which, though some deem it not inconsistent with Theism, is undeniably the pet and toast and boast, the philosopher and orator and household gods, of the latest atheism trying hard to explain the natural without help from the supernatural. This time, also, the challenge is many-voiced, coming as it does from all the many *Revolutions* of the System. Instead of falling down before an atom (I am quite willing to say a *universe* of atoms, if you choose) and saying, Thou art my Maker, the Revolutions prefer to go further for their worship, in hope of faring better ; and send off from all their westerly, elliptic, and inclined orbits unlimited tangents and centrifugals of telegram into the great void, asking, Where is He ? For *that* He is, they can have no manner of doubt. Where the Law Scheme does not answer God must be accepted. There is no *tertium quid.* As science now stands, the only alternative to eternal evolution is an eternal Creator. Who to-day believes in chance ? Newtons and Miltons, celestial systems by hosts, Principias and Iliads, made by a fortuitous concourse of atoms — who now believes that, or is in any danger of believing it ? Law, eternal law — this is the present chorus of all unbelievers. And when law breaks down, what but a personal Creator can the reasonable thought fall back upon? Nothing. It goes up and down the spaces asking for God. Nor does it ask in vain. Though azure seas on seas may say, He is not in us; and deeps on deeps beyond may say, He is not in us; and a still remoter hell and destruction may say, We have only heard the fame thereof with our ears — He is at last found sitting in the very zenith, and on the circle of the heavens. And then the whole Solar System, from center to circumference unexplained by mere naturalism, and from center to circumference explained by the supernatural, solemnly lifts confessing hands, thick as grain stalks by the Nile, and says, *I believe in God, the Father Almighty, Maker of heaven and earth.*

This ends my list of facts from the Home Field. It is a small field — only some thousands of millions of miles across, and ruled by a globe only about a million of miles in diameter — but then it is near to us, and we are able to see things in detail to an extent impossible in remoter regions. And, altogether, for so narrow a district, it has quite a breadth of story to tell about the Nebular Hypothesis.

In the next lecture we will pass on to a wider domain. We will pass from the Home Astronomy to the Foreign, from the canton to the empire, from our Solar System to that distant realm of stellar and nebular glory where distance is the least of the things that "lend enchantment to the view." Here, if I mistake not, we shall hear the same testimony from richer voices, and from the very latest Astronomy. Sun-clouds, and clouds of suns, will "take up the wondrous tale and repeat the story of their birth." And it will be the story of a birth, not by law, but by God. Perhaps, these higher witnesses will be even more communicative than those we have just heard. Perhaps, even, we shall find their gleanings better than the vintage of the Solar System. This System, by the present extreme light and heat of its center, by its various chemical constitution, by its diverse mechanical relations, and by numerous features of its rotations and revolutions, has already told us much. And it might have told us more if we had chosen to cross-question it — especially if we had chosen to ask, not merely about things inconsistent with the Nebular Hypothesis, but also about things which it leaves unexplained.

But the most will be told by that great *Foreign* Realm whose breadth fatigues our thought: and, as we pass along its glorious highways, systems after systems will present themselves to us, and almost *ask* permission to drown with their sublimer, but chording, voices the witnessing of the Solar System. Hear! "Lo, the heavens are not self-sown. Their bright harvests are not of spontaneous growth. Yonder great prairies of the sky did not clothe themselves in the green of suns and worlds — did not stock themselves with these astonishing conservatories, and rear amid them these gleaming Sydenhams of beauty and wonder. Never did such palaces build their own splendors — never did such gardens do their own sowing and planting and arranging. There is a Heavenly Sower, Planter, Builder. Some supreme Virgil sung these wonderful Bucolics in worlds. Some celestial Linneus set up this celestial Jardin des Plantes. Some Divine Person and Potentate set up these magnificent Concordats through the heavens, far and near."

So, with voice that ought to make itself heard specially by every astronomer, and well by every one who can understand astronomy, the whole astronomic field, domestic and foreign, will declare itself against that scheme of naturalism which not only makes the celestial systems begin and end in smoke, but finds in that smoke all the attributes of a Creator.

VIII.

CONFLICT WITH STELLAR ASTRONOMY.

Εἶς ταῖς ἀλεθείαισιν, εἶς ἐστιν Θεὸς,
Ὃς οὐρανόν τ᾽ ἔτευξε. — *Sophocles.*

Bene autem universus mundus Dei templum vocatur, propter illos, qui æstimant nihil esse aliud Deum, nisi cœlum et cœlestia ista quæ cernimus. — *Macrobius.*

VIII. Conflict with Stellar Astronomy.

1. A PRINCIPLE 246
2. VISIBLE SYSTEMS 248
3. NO HUGE CENTERS 250
4. WITHOUT CERTAIN GRADUATIONS 252
5. VARIOUS PLANES 256
6. ECCENTRIC ORBITS 258
7. DIFFERENT CHEMISTRIES 259
8. A DREAM 262

EIGHTH LECTURE.

CONFLICT WITH STELLAR ASTRONOMY.

TRUE science is not easily overvalued. Only one thing is worth more. What that is you do not need to have me say in so many words.

As much cannot be said in favor of what is sometimes called scientific speculation. That is a very different matter. While often useful, and even necessary, as preparing the way for science, it is not seldom totally worthless and even pernicious in its scope. It would be hard to find among the professed rhapsodies of poets things more extravagant in conception, more lame in argument, and more strange to the world of fact and experience than are some of the notions now being put before the world under the great name of science. The Doctrine of Metempsychosis is a great piece of sobriety compared with the Doctrine of Evolution.

In the last lecture, I spoke of the origin of worlds and astronomic systems; and tried to make clear that the actual Solar System offers to the Nebular Hypothesis insuperable objections. Among these objections are the present extreme light and heat of the sun, the various chemistry of the System, its diverse mechanical relations, and whole sheaves of difficulties bound up into two by the words *rotations* and *revolutions*.

We will now widen our view. Instead of a horizon which just manages to pass around Neptune and the most outpost comet, let us have that grander horizon which takes in THE FIXED STARS WITH THEIR VARIOUS SYSTEMS OF KINDRED SUNS.

What say these higher systems to the Nebular Hypothesis — to the astronomical part of that Law Scheme which is now so generally used in the interest of unbelief, and on which so many are now gliding, with all sails set, into the black deep of atheism? Do they speak against it? If so, what do they speak?

Before answering these questions, I must call your attention more particularly to the full meaning of the Nebular Hypothesis. It means much more than appears on its face. It means that each stellar system, however large, is derived from a single fire mist: for, if it supposes several fire mists as the source of the system, it has to suppose them existing as a system of revolving bodies, and so begins its explanation of Nature at a point which itself needs explanation. Also, the scheme cannot avoid calling in the supernatural save by supposing that each group of worlds, built up after its manner, finally falls together at its center of gravity and goes back to the gaseous state, and so on in eternal cycle: and the same natural causes that would bring this about for each planetary group belonging to a stellar system, would bring it about for the whole system however large, and from time to time resolve it into one fire mist. Accordingly, it is an essential part of the Nebular Hypothesis, as held by atheists, that not only such a small system as our sun presides over, but also all those group and cluster sun-systems whose vastness appals the imagination, and even that ultimate system which throws its stupendous tentaculæ about all the starry nations — that each of these came from one great nebula. Just think of that greatest nebula of all, that one all-comprehending fire mist, that Mighty Cloud which at some remote time was the sum of all material things! It is from this last unit of

Astronomy — contracting, rotating, accelerating, casting off rings, compacting its rings into spheres — we must manage to get such stellar systems as we find peopling the remoter depths of the heavens. And, in general, each system of stars, whatever its size, must have come from one nebula ; and this one nebula, as we have already seen, could only separate itself into worlds by the breaking up of successive equatorial rings.

Now let us notice some facts as to the stellar systems inconsistent with these views.

1. *There are multitudes of stellar systems distinctly and gloriously visible — not a few of them of immense size.*

According to the hypothesis, there ought to be no *visible* self-luminous stellar systems at all ; at least none of many members. Judging from our own solar group, by the time the central part of a fire mist is in the state of our sun, all the bodies thrown off from it ought to be cool and without light of their own. Even Mercury shines only by reflected light. It has cooled away into darkness before a new zone has been cast off from the sun, or the solar surface ceased to be at white heat. Now the central parts of many systems appear quite like our sun ; and so the systems should have no visible distant outskirts — every system ought to consist of, at most, only one or two visible members. But, in reality, we have shining on the naked eye, and especially on the eye of the telescope, hosts of systems much larger than this: sometimes of prodigious size both as to number of orbs and the space through which they are distributed; wholly self-luminous, as their spectra show; and not seldom showing on their remote borders as intense a brilliancy as at the very center.

2. *Many a stellar system is without a dominant central orb.*

Though there is often considerable difference in size among the members of a star-group, yet there is seldom, if ever, so great difference in favor of some one star as the analogy of our system and the principle of the Nebular Hypothesis seem to call for. In some cases what seems the heart of the system is held by a body no larger than any other member; in some cases it is held by a body much smaller than the average; and in by far the greater number of cases it is held by no body at all. The center of gravity is in mere empty space. The real pivot of the system is totally invisible. Most astronomers would say with Humboldt that this is true of all the multiple stars. Like too many of us, they revolve about nothing.

But this is not according to the hypothesis — which requires at the center of *every* system, not only some orb (for I would like to know how it is *possible* for a fire mist, by mere rotation, to empty its center of all matter), but also an orb much greater than any other member of the system, and great in proportion to the size of the system. In all the satellite systems which we can observe, the center is held by a body not only much larger than any one companion, but even much larger than all its companions put together. The same is true of our Solar System. In both stature and governing power the sun is overwhelmingly the king of the group. Much more ought there to be a kingly visible center to all those much larger stellar systems which show themselves in remote space. For the larger the fire mist, out of which a system is made, the greater, other things being equal, must be the density of its central region, and the greater the centripetal force at a given distance from the center, and of course the greater the distance from the center at which a given centrifugal force will succeed in casting off a ring. Also, the denser the central region is the less will it contract by a given loss of heat, and the sooner it will reach the point where it will not contract at all. Thus the greater central density of a large nebula must act in two ways to give a greater central orb, namely, by increasing the centripetal force, and by resisting the growth of the centrifugal by contraction. Hence, as the very smallest systems, and all which are known to us, have dominating central orbs, much more will the great stellar systems have them : and the larger the system the larger will be the orb. This is according to what we observe among the small systems with which we are connected ; which range in the following order, both as to the size of the system and the size of the central body — the Earth's, Neptune's, Uranus's, Saturn's, Jupiter's, the Sun's. How amazingly large ought to be the central world of such a system as the great cluster in Hercules, or the Milky Way! It ought to be an emperor. It ought to be Cæsar Augustus among the emperors. Alcyone, especially, ought to appear in our sky with almost solar glory. Though now some 12,000 times larger than our sun, it ought to be millions on millions of times larger still. And our benighted earth, whose sky is perpetually illustrious with mighty groups, ought to be so far like that Better Country of which we have heard, and which we reverently hope for, that we could say of it, And there is no night there.

3. *Many a stellar system presents no graduated appearance as to the light, distance, and size of its members, from the center outward.*

According to the Nebular Hypothesis, the stars in each system were ripened successively, at immense intervals, and are in widely different stages of combustion. These different stages ought to show themselves in a certain graduated aspect of the system as to light. Its brightest part should be the center, and it should gradually shade away toward the outskirts. This should be specially noticeable in large systems.

That there is a considerable number of systems whose light is graduated after this manner is well known. The trouble is that they are not universal. Nay, the trouble is that they make but a small part of the stellar domain. Almost all the multiple stars, and scattered groups consisting of members physically connected, may be cited in proof. In these, whatever star you may take as the structural center of the system, you cannot make out a gradual fading in the light as the eye passes outward from star to star. The same is true of some large clusters. In nearly four hundred systems out of six hundred, as examined by Struve, the stars throughout are of the same color and intensity of light; in others the brightest and whitest of the stars are at the outside of the group; in still more cases the different sorts of stars as to light are wholly intermingled as if at random. The Pleiades, as shown by the telescope, are an example of this last class. And the examples are comparatively very few of systems whose central glory steadily fades and dies away toward the suburbs — like almost every ancient city, or like that famous ancient empire which had Augustan Rome at its center and the rude Britons and Goths and Arabs at its circumference. Instead of finding here and there a case of this sort, or here and there a considerable number of cases, we ought to find absolutely no others. The Nebular Hypothesis being witness.

In regard to the size of the members of a system and the distance between them — these ought to increase steadily from the center outward. For, in the shrinkage of the mist, the centrifugal force must increase faster than the radius diminishes ; so that, the nearer the center, the more frequently rings would be thrown off and orbs formed. Also, the nearer the center, the smaller would the ring be, both in diameter and in breadth of actual matter; and so the less matter would it contain — notwithstanding the increase of density toward the center, under the influence of gravity. In fine, it would be in all systems as it tends to be in our own — the larger bodies and intervals at the greater distances from the center. As is the way of suburbs, the remotest structures would be the furthest apart: as is not the way of suburbs, the remotest structures would be the largest of all. The elder children of the family would be the largest and most independent — as in a natural scheme they ought to be. The stoutest soldiers would guard the perilous frontiers — as in a natural scheme they ought to do. In the case of very large groups and clusters, the outpost worlds and intervals would be enormous compared with the rest. Such a cluster as the Milky Way, that City of magnificent distances and sizes, ought to show on its frontiers distances and sizes incomparably most magnificent of all. The very giants of the system should be there. Tellus, Typhon, Enceladus — the very Olympians are afraid and fly to Egypt. And each giant mounts guard over a district proportioned in extent to his own sublime stature.

How different all this is from our actual Astronomy, every observer of the heavens knows. Perhaps, one might, with much pains, hunt up a few stellar systems which make a show of conforming to these views. But they are very few. Most, to say the least, of the physically connected multiple stars and scattered groups are plainly of quite another stamp. They have no such graduation as the hypothesis requires. By far the greater part have no graduation at all: their various sizes and intervals are scattered about as by some celestial lottery. And, in some cases, we find exactly the opposite of what we are taught to expect. There is a general tenor of orderly arrangement; but instead of proceeding outward from the less to the greater, it proceeds from the greater to the less. The smallest worlds and intervals are at the outside of the system. And one of the most striking examples of this would seem to be given by that very Milky Way which, on account of its hugeness, should be among the very last systems to give it. Sir John Herschel was strongly drawn, by his extended observations within our cluster, to the opinion that the stars in its outer parts are generally really smaller and more densely placed than the others.

4. *The stars of the same system are often in different planes, while they are not known to be in the same plane in a single instance.*

I have already called your attention to the fact that the bodies of our Solar System move in different planes. A similar fact is written quite as plainly, and far more strikingly, on some of those larger systems of which I am now speaking. On taking the inclination of the visual ray to the orbits of the double stars — all of which belong to our Milky Way system — we find the angle exceedingly various. On criticising the relative motions in some multiple stars and other small groups, we find them explainable only on the supposition that the orbits of the same sub-system, as well as of different sub-systems, in our cluster, are largely inclined to each other. The general telescopic aspect of some still larger systems tells the same story of them; for, they are so densely crowded toward the center, and are otherwise so characterized, as to force on us the idea of stars arranged in globular or other solid forms. The great clusters in Hercules and Libra, and that known as 30 Doradûs, may be taken as examples. Of course, in a globular cluster the orbits run through the whole gamut of inclinations. They stand out from each other like the spokes of a wheel, or rather like the miscellany of great circles forming a skeleton celestial sphere: they bristle away from each other as if charged with electric repulsions: one half get as far away from the other half as they possibly can. No frisky, impetuous comets can abhor each other more cordially, cut each other more unkindly, object to their ecliptic more strongly, than do hosts of staid and massive suns bound up together in the various families, clans, and celestial nations of the same globular cluster.

This, on the one hand. On the other, we do not as yet know a single stellar system whose orbits lie exactly in the same plane. How is this reconcilable with the notion that the worlds in each system have sprung from one central rotation? Our Milky Way is a system by itself. If all its stars were in one plane with ourselves we should see them all projected on a great circle of the sphere in one narrow but most brilliant band of light. Such a cestus, fairer and more marvelous than poets ever gave to Venus, we do not see. On the contrary, we see the stars of our cluster scattered all over the sky, and so know that they occupy innumerable and widely different planes. According to the theory, they all ought to be found in the ecliptic, and make a *visible* ecliptic, bright as the electric arc, across the heavens. Nay, all clusters and nebulæ whatsoever ought to lie on the same circle: for, they must all be supposed to have sprung from one monster fire mist whose one rotation was along the ecliptic. But, in fact, there are immense nebular accumulations at the very poles of the Milky Way. Those arctics of Nature seem bedded in eternal snows. They are twin breakers around which the celestial seas are always breaking in clouds of silver foam.

5. *The stellar systems, as far as examined, show very eccentric orbits.*

I have already given reasons for claiming that worlds formed in the manner of the Nebular Hypothesis must revolve in circles.

But the orbits of the double stars which have been completely made out, amounting to about a score, are very like those of comets. In two cases, those of Alpha Centauri and Gamma Virginis, the orbit is nearly five times longer than it is broad ; and generally the length exceeds the breadth by more than a quarter of itself. The experience of astronomers in this direction is so uniform, that now, whenever one sets himself to find the elements of a new sun-orbit, he expects, if successful, to see it turn out very eccentric, and would not be at all surprised to find it as sharp and oval as the comet of Halley traverses — that is to say, almost a celestial needle, with the sun for its eye, piercing the night. — About seven hundred double stars have shown more or less orbital motion. Such has been the character of the arcs thus far described that it can hardly be doubted that, should time and pains enable us to extend and close up all these arcs into orbits, we will find very few of them to be accurate circles. Circular rings are as scarce in the heavens as elliptical rings are in our boxes of jewelry.

6. *Many a stellar system has not the same chemical constitution throughout.*

We have found a like fact nearer home. Our own solar group is far from being a unit in constitution. Especially, when we compare the earth with the sun, and both with the comets, are we struck with the difference in this respect.

A difference not compatible with the idea that sun and planets and comets are but different specimens of the same well-mixed and homogeneous fire mist. So I have already attempted to show.

But now let us turn to those far off systems which, until lately, seemed far enough beyond the reach of our chemical critics. Of what are they made? Are the members of the same system always made of the same materials, in the same general proportions? We are not without an answer that can be trusted. Many double and multiple stars are found composed, each of differently colored members, whose difference of color can hardly be supposed due to contrast, or to different stages of combustion. Besides, when probed by the spectroscope, these variously colored stars give forth various spectra — showing that they have different chemical constitutions. Indeed, stars of the same system seldom give precisely similar spectra: while in some cases the difference is very great and radical. Thus, in our cluster, Betelgeuse and Beta Pegasi and Alpha Orionis send out no lines whatever of hydrogen, an element found so largely in our sun. And our sun gives no sign of oxygen or nitrogen — elements found largely in some other parts of the great stellar system to which it belongs. Sirius has strong rulings through the violet which do not answer to any known substance: and, on being put well through the spectral catechism, confesses to many points of singularity, especially in regard to the proportion of the elements in its great seething alembic. The lines of hydrogen are far stronger than those which come from our sun, while the metallic lines are far fainter. If these large bodies were all parceled off from the same fiercely boiling and thoroughly mixed nebula, they could not show such varieties of constitution as we are able to detect by peering through the grated windows of their spectra.

This ends my list of facts from the confessed stellar systems. Please set each of them down as against the latest *atheism* — since each is against that only scheme of naturalism that now attempts to build the heavens without a God. And remember that the attacking force of these facts is measured, not by their sum, but by their *product.*

Putting, then, our facts together geometrically, one feels that if he could only leave this distant post of observation from which he yet sees so much, and, harnessing twin stars to some heavenly chariot, could ride freely in and out among the stellar universes, he would find almost endless incongruities between them and the idea that their elements came together of themselves, in however much of time, (call it the brother of eternity) into such a Glorious House — as bewildering in its elaborateness and unity as it is in its vastness — as the German poet saw with his heart that wept and trembled.

"God called up from dreams a man into the vestibule of heaven, saying, 'Come thou hither, and see the glory of my house.' And to the servants that stood around his throne He said, 'Take him, and undress him from his robes of flesh: cleanse his vision, and put a new breath into his nostrils: only touch not with any change his human heart — the heart that weeps and trembles.' It was done: and, with a mighty angel for his guide, the man stood ready for his infinite voyage; and from the terraces of heaven, without sound or farewell, at once they wheeled away into endless space. Sometimes with the solemn flight of angel wing they fled through Zaarrahs of darkness, through wildernesses of death, that divided the worlds of life; sometimes they swept over frontiers, that were quickening under prophetic motions from God. Then, from a distance that is counted only in heaven, light dawned for a time through a sleepy film; by unutterable pace the light swept to *them*, they by unutterable pace to the light. In a moment the rushing of planets was upon them: in a moment the blazing of suns was around them.

"Then came eternities of twilight, that revealed, but were not revealed. On the right hand and on the left towered mighty constellations, that by self-repetitions and answers from afar, that by counter-positions built up triumphal gates, whose architraves, whose archways — horizontal, upright — rested, rose — at altitude by spans — that seemed ghostly from infinitude. Without measure were the architraves, past number were the archways, beyond memory the gates. Within were stairs that scaled the eternities below; above was below — below was above, to the man stripped of gravitating body: depth was swallowed up in hight insurmountable, hight was swallowed up in depth unfathomable. Suddenly, as thus they rode from infinite to infinite, suddenly, as thus they tilted over abysmal worlds, a mighty cry arose — that systems more mysterious, that worlds more billowy — other hights and other depths — were coming, were nearing, were at hand.

"Then the man sighed, and stopped, shuddered, and wept. His overladened heart uttered itself in tears; and he said — 'Angel, I will go no farther. For the spirit of man acheth with this infinity. Insufferable is the glory of God. Let me lie down in the grave and hide me from the persecution of the infinite; for end, I see, there is none.' And from all the listening stars that shone around issued a choral voice, 'The man speaks truly: end there is none, that ever yet we heard of.' 'End is there none?' the angel solemnly demanded: 'Is there indeed no end? — and is this the sorrow that kills you?' But no voice answered, that he might answer himself. Then the angel threw up his glorious hands to the heaven of heavens, saying, 'End is there none to the universe of God. Lo! also there is no beginning.'"

IX.

CONFLICT WITH NEBULAR ASTRONOMY.

Χάος ἦν, καὶ νύξ, ἔρεβός τε μέλαν πρῶτον, καὶ Τάρταρος εὐρύς. Γῆ δ', οὐδ' ἀὴρ, οὐδ' οὐρανὸς ἦν· ἐρέβους δ' ἐν ἀπείροσι κόλποις τίκτει πρώτιστον ὑπηνέμιον νὺξ ἡ μελανόπτερος ὠόν. — *Aristophanes.*

Quis credat tantas operum sine numine moles,
Ex minimis cœcoque creatum fœdere mundum ?
Manilius.

IX. Conflict with Nebular Astronomy.

1. NEBULÆ 267
2. EXAMINED BY SPECTROSCOPE 269
3. SHOWN TO BE FIRE MISTS? 270
4. SHOWN NOT TO BE 273
5. WHAT IF THEY ARE? 285
6. THE WHOLE FIELD 291
7. A VOYAGE 297
8. THE WHENCE AND THE WHITHER 300

NINTH LECTURE.

CONFLICT WITH NEBULAR ASTRONOMY.

IN searching for facts bearing on the Nebular Hypothesis, let us now widen our already wide horizon so as to take in those celestial clouds and cloudlets, which, under the name of nebulæ, have not yet been separated by the telescope into ordinary stars, but which many of us have been wont to regard as only remoter star-crowds converted into a silver haze by extreme distance from our post of observation.

The nebulæ — what are they ? They *seem* like silver smokes ; beds of eider-down ; faint celestial frosts ; milky ways ; snowy plumes shorn from the crests of heavenly warriors ; delicate white foams such as the goddess Venus was developed from, according to the ancient fable, and such as the planet Venus was developed from, according to the modern fable. But what are they really ?

The chief support of the Nebular Hypothesis, in

these days, lies in the supposed evidence that the nebulæ, in part at least, are specimens of such fire mists as would be needed for the natural construction of worlds. If, on examination, it shall appear, not only that there is no such evidence, but that there is positive evidence to the contrary, the main support of the Nebular Scheme will be set aside in the fullest manner.

For the sake of clearness, I will here lay down three propositions to be examined successively in the light of the more recent Astronomy.

1. *There is no positive evidence that any of the nebulæ are fire mists.*

2. *There is positive evidence to the contrary.*

3. *Granting that some nebulæ are fire mists, it is plain that they are not such fire mists as the Nebular Hypothesis demands.*

Let us see how far these statements are borne out by facts: and, first, the statement that there is no positive evidence that any of the nebulæ are fire mists.

The spectroscope is found to give for some of the nebulæ those bright-lined spectra generally given by gases in a state of incandescence. Hence it is concluded that these nebulæ are fire mists. I think without sufficient reason. To lay no stress on such perplexing facts as that the solid erbium gives the gaseous spectrum, that the spectrum of a substance is often greatly altered by chemical combinations and sometimes totally suppressed, that gases (hydrogen for example) under high pressure and temperature sometimes give the continuous spectra of solids — the most these facts show is that certain nebulæ are, to some extent, gaseous in composition. They by no means show that each nebula is one continuous fire mist — any more than the continuous spectra found belonging to certain other nebulæ, as that in Andromeda, show that they are so many continuous solids. This no one supposes. It only is supposed that such nebulæ consist of distinct stars largely solid. So the nebulæ with the bright bands may as reasonably be thought to consist of distinct stars largely gaseous. Indeed, some well-known separate stars, and clusters of stars, are found to give the bright-lined spectrum — for example, three stars in Cygnus, the Dumb Bell Nebula in Vulpecula, the Crab Nebula in Taurus, the Ring Nebula in Lyra, as well as the irregular Nebula in Orion ; each of which has been largely resolved into stars, and shows throughout the characteristics of resolvability strongly marked.

There seems to be no way of keeping the peace between the telescope and the spectroscope, in regard to such nebulæ, except by allowing them to be *clusters of at least largely gaseous stars.*

But this is not all. A nebula with the gaseous spectrum may even consist of discrete stars *very much like our sun* — assuming, what is still in dispute, that our sun is partly gaseous and partly solid. For, if a nebula is made up of separate suns, each of which has a solid nucleus surrounded by an incandescent atmosphere so deep and dense and various in its elements as to virtually suppress the light from the interior solid, then a gaseous spectrum would be given if there were still another incandescent atmosphere outside of each orb. Thus, suppose an incandescent solid. Envelop it in glowing gas consisting of only one element and you will have the spectrum crossed by certain dark lines. Every new element added to that atmosphere in sufficient quantity would add to the number of such lines. Of course, you can conceive of the number becoming so great as to make the spectrum very faint, and even null. If now there were placed around that glowing atmosphere still another, we should get from the whole a gaseous spectrum.

And so we should if each sun has a solid nucleus very much less bright *to us* than its gaseous envelope — or, if the two are separated by a dense non-conducting stratum of some sort, such as some still suppose may protect and make habitable the body of our sun. If a calcium light at its fiercest is placed between the eye and the sun, it appears as a black spot on the disc from the effect of contrast. So when the bright lines which make the gaseous spectrum are thrown across a much brighter continuous spectrum they appear as dark lines — if across one of about the same brightness they do not appear at all; nothing but the continuous spectrum is seen. But if they are thrown across a much dimmer continuous spectrum they must still show as bright lines, and may make the whole underlying spectrum very faint and even invisible. So the most we are at liberty to conclude from the fact that a nebula gives the gaseous spectrum is that whatever solid matter belongs to it is somewhat less bright than its envelope to the eye. This does not necessarily mean that it is less heated, or even intrinsically less luminous. It may mean only that it is chiefly more distant from us — say in the orbs at the heart of the nebula — and that much light is lost in traversing the radius of the nebula.

But, really, it is only a part of the truth to say that some of the nebula give the gaseous spectrum. They do give that, but in many cases, to say the least, they also give in the background a faint continuous spectrum, or the suspicion of one, overpowered by the superior brightness of the gaseous spectrum. And the delicacy of this sort of observation is so extreme, and the course of experience has been such, that no one has a right to say that the continuous spectrum does not really lurk beneath every nebular spectrum. Repeatedly, a nebula supposed to give only bright bands has, on more careful scrutiny, or use of superior instruments, been found giving a faint satellite spectrum with its continuous ribbon. The great nebula in Orion is a striking example. Till recently this object was habitually pronounced with great confidence to be, throughout, purely gaseous in its message to us ; but at last we have detected a confused tartan hiding beneath the green and blue bars of nitrogen and hydrogen.

So it is very far from being clear on spectroscopic grounds that any nebula is mere fire mist. And the intrinsic difficulties in the way of accepting such a view are by no means small. This brings me to the second proposition which we set out to examine, namely, There is positive evidence to the contrary.

First, *It is hard to understand how a mere fire mist, of the utmost tenuity, could be seen at such distances from us as most of the nebulæ evidently are.*

They have no sensible parallax. But if any star, or cluster of stars, at the point of sensible parallax, were diffused through millions of times its present space it would become totally invisible to the most powerful telescopes — unless its intrinsic brightness were at the same time millions of times increased. But the stellar brightness, if we may judge from that of our sun, is already intrinsically intenser than that of the most powerful electricity known to us. Whoever assumes a gaseous light millions of times brighter than even this, travels away infinitely from the region of experience ; a thing which naturalism is not allowed to do. Never, among the host of incandescent gases which under the name of comets steam through our planetary seas, have we seen a specimen of such ineffable brightness, or indeed of any not vastly inferior to that of our sun.

But this is not all. Experiment also has a word to say. It is found that a high temperature gives to nitrogen a spectrum of many rich bands. As we reduce the heat the number of bands is reduced. At last, when incandescence is feeblest, but a single member of the gay specterhood is left. Now it is this member clad in Kendal green, and this alone of all those in the livery of nitrogen, that appears in the spectra of all the supposed fire mists. The inference is that their temperature is very low. Are such bodies hot and bright enough to report themselves to our sight from almost the outskirts of Nature?

Suppose the most brilliant comet on record carried away as far as the great nebula in Orion, and then pieced out with cometary matter of equal brightness till as large as that nebula, who supposes it could then be seen, even in that Light-Compeller, the great Rossian speculum? And yet the Orion-Nebula is just visible to the naked eye.

One can understand how gases under vast pressure, as at the surface of sun or star, can be about equal to solids in brightness: but it is only the central part of a fire mist which can give gases in such extreme density.

Second, It is hard to understand how mere fire mists can have, or seem to have, any other permanent shape than the spheroidal.

The spiral nebulæ, some forty thus far known, firmly maintain their strange and exceedingly various general shapes. We do not wonder at this. These bodies, giving continuous spectra, are admitted to consist of distinct solid orbs; and we can easily conceive that a system of such orbs may have its motions, distances, sizes, densities, and brightnesses so adjusted as to maintain for long periods almost any apparent figure. Indeed, it can be shown that these elements may be so related to each other that almost any apparent shape might be given by what is really a perfectly globular system of worlds. But how can a continuous vapor out in free space maintain itself for ages in any other one shape than a spheroidal? It were to defy all known laws of gravity and equilibrium. These laws which on the earth strain all falling fluids into globules — which are supposed to have rounded all the solid members of our solar system, all the separate stars, and nine tenths of the nebulæ — and which gradually compose into spheres even the most misshapen comets as they get away from the fiery repulsions of the sun — these laws would incontinently compel any irregular aeriform body, placed all alone in space, toward a spherical form. As fast as the mountains would run down to one level if suddenly changed to water ; nay, as fast as the mountains of hydrogen at the sun's edge are seen to disappear ; so fast would the outline of the most irregular fire mist tend to straighten itself out into an exact circle or ellipse. No forces wholly within itself could sensibly defeat this tendency. Jets might be thrown out from the general sphericity ; but, from the nature of the case, they would be exceedingly transient and shifting, as well as small compared with the diameter of the whole nebula. They would also be the thinnest and least evident part of the whole. And, at such immense distance, they would not appreciably break the general roundness of the nebular figure to the eye. Just as the earth's atmosphere keeps to its general roundness though subject to violent local heats and storms ; just as the sun shows a round face though it has great centers of commotion and outburst ; just as the planets and their moons give us circular discs though no doubt ragged with mountains, and though — such is the hypothesis — born and nursed and settled in great gaseous tempests ; so all gaseous nebulæ, however disturbed from within, ought to appear round to us. But many give no sign of being disturbed from within. They have a very uniform look. We can detect no nuclei and centers of violent action. Even such nebulæ, as well as others, often appear under the most irregular shapes — as various as those of the clouds — and keep these shapes without perceptible change. The neighborhood of the Milky Way is rich in such nebulæ. The famous Dumb Bell in Vulpecula is as much a dumb bell to-day as it was at its discovery.

But some one says, " The apparent figure of a nebula is not of course the true one. By enlarging the power of the telescope we often greatly alter the nebular figure, by bringing into view parts too faint before to be seen. Thus the giant Rossian tube almost seemed to create the Crab Nebula and the great Spiral. And who can say that the increase of power might not be carried so far as to reduce all the nebulæ to roundness? Suppose a globular fire mist with its outer part too faint to be seen, and with very unequal density and brightness at the part where it becomes visible — then it would appear very irregular in form. Why may not all the irregular nebulæ be fire mists of this sort? "

I answer that at least some of them show traits inconsistent with this view. Some show the same outline whatever the telescopic power used. Others are as sharply defined as a new coin. Still others are of uniform brightness throughout. And some have all these traits permanently — for example, the two nebulæ appearing, one as a spectral scimitar, and the other as the spectral hand of Saladin stretched out to grasp it. Such nebulæ can hardly be globular. But if they are, it is as much against the law of equilibrium for any part of the interior of a fire mist to remain permanently in one irregular shape as it is for the whole mist to do so. Thus, whether we regard the apparent outline of an irregular nebula as the real or not, its constancy is against the hypothesis that it is a fire mist.

It is worthy of note that increasing the power of a telescope is apt to increase the irregularity of a nebular outline. Thus it has proved in the case of the annular and elliptical nebulæ — and thus it has memorably proved in the case of the most striking Whirlpool Nebula, which in a common telescope appears as two globular mists, but at Parsonstown blazes out into an elaborate and most magnificent spiral maelstrom of light.

Third, It is hard to understand how mere fire mists can retain for long periods such an internal configuration as some of the gaseous nebulæ, so called, show.

Here are several annular nebulæ giving the gaseous spectra. Such rings once formed would be stable if they should revolve at such a rate as to make the centrifugal force equal to the centripetal. But how could such rings ever *get* formed? How could the mere revolution of a continuous gas empty its central region entirely of matter?

Here are certain irregular nebulæ giving gaseous spectra. Within them we find large vacancies; often very sharply defined, and irregular in outline, and surrounded by a very equable luminousness. How such vacancies can occur in a fire mist at all — especially, how they can maintain their size and shape and place unchanged to the eye in all respects from generation to generation is not easily made out. It is conceivable that a mass of distinct *moving* stars may show such aspects; because it is conceivable that such a mass may remain stable for long periods in almost any form. Not so with a fire mist of amazing mobility, in the most unsettled stage of its history.

Here is a nebula, pronounced gaseous by the

spectroscope, which has a ridged and mottled face, or a nucleus at its very edge, or several nuclei remote from a center which shows no sign of special condensation — these features remaining without perceptible change ever since that distant time when they were first observed. Such fixity in the midst of a raging and billowy sea of vapor — to say nothing of the difficulty of reconciling such a disposition of nuclei with the fundamental conception of the Nebular Hypothesis and the laws of equilibrium — seems incredible.

Fourth, It is hard to understand how mere fire mists can alter their size and brightness so slowly, if at all, as the gaseous nebulæ are seen to do.

Great and sudden changes have been reported in the nebula about Eta Argûs. That in the sword of Orion has been suspected of changing in both place and form. Several other nebulæ, it is thought, have varied in light and even disappeared. But there can be no doubt whatever that there are many nebulæ, pronounced gaseous, in which the most careful observation, continued for many years, has failed to detect any change in main features ; especially a steady loss of light and size. They are stereotyped. They stand today as they stood in the telescope of the elder

Herschel, and of Messier, and of Huyghens: the Milky Way, with all its smoky spurs of gaseous autograph, we have reason to believe, appears to our eyes as it did to the eyes of Hipparchus and Ptolemy.

Now this, while easy of explanation on the supposition that these nebulæ consist of discrete stars, is very hard of explanation on the supposition that they are mere fire mists. There is no body with which we are acquainted, in all our dealings with heated bodies on the earth, but will, if left to itself, cool down to the general temperature of surrounding space in a few hours, at furthest. In the case of aeriform bodies the cooling is particularly rapid — especially when their temperature is very high and that of surrounding space very low. It means much — that ancient saying from a book whose very poetry is sometimes science, They shall vanish away like smoke. The glowing gases of our laboratories disappear like a flash at the touch of the winter air. Incandescent comets totally pass from observation at a very small remove from the sun. Now, each fire mist is supposed to be at a terrible heat, and the immediately surrounding region is known to be terribly cold. Such a mighty hunger and such a congenial and convenient repast could not be kept apart. Space at —132° Fahrenheit, would drink off the heat of such a gaseous body (with its enormous faculty of transmission and convection) at a fearful rate — as no Sahara ever could drink water — especially at the earlier stages of the mist, especially at its outside, especially from detached wisps and streams and protuberant equator, especially from such protuberant equator when become very flat and separated from the main body as a ring. Such rings and wisps are admirably fitted for cooling. Heat would flow off from them with miraculous fluency. They would empty themselves into the void with unspeakable precipitation. Condensation and loss of light would proceed equally fast. The aspect of the nebula to us ought to change almost daily. In the course of a few weeks or months, at the most, we ought to see its most exposed parts ripen swiftly toward rings, planets, satellites. And yet even the annular nebulæ, and those whose outline and internal structure are of the most broken character, for example 17 Messier, show no change whatever in size or light as great periods elapse.

If one says that conduction of heat in a gas is very feeble, and that this would keep up the heat of the interior, I answer that this would make the cooling of the outside of the fire mist all the more rapid.

If one says that the same reasoning would require the sun and stars, which are known to be fire balls at least partly gaseous, to change perceptibly in size, brightness, and structural appearance within short periods, whereas no such changes are observed — I answer that these bodies are so enormously inferior to the supposed fire mists in tenuity, mobility, amount of exposed surface, as well as in the violence and disorder of internal forces, that the changes within a given time in size and brightness and structural aspect ought to be enormously inferior also. Loss of heat in a given time would be less, and a given loss of heat would produce less effect on the volume and brilliancy of such a body as the sun than it would in the case of its less dense and more heated corresponding fire mist. That fiery Cloud — glowing with a heat beyond all imagination, and assailed on all sides by millions of huge, insatiable Siberian mouths draining off its fiery wine in so many impetuous gulf-streams — must change almost like the clouds in our fickle terrestrial sky.

Fifth, It is hard to understand how mere fire mists can be sharply defined and show uniform brightness throughout.

When such a mist has ceased to expand, its outside cannot be very hot, because then the heat and gravity are in equilibrium, while the gravity must be very small on account of the very great distance from the center. Hence every fire mist ought to appear very dim at its edge; and, since the heat and thickness continually increase toward the center, ought to brighten gradually in that direction. Besides, the law of gravity requires special condensation at the center. But, in fact, many of the gaseous nebulæ, so called, instead of showing a thick creamy center and diluted margin, show a sharp definition and a uniform brightness throughout. This is true of most of the planetary nebulæ — all of which, as far as examined, give the bright lined spectrum. It is also true of very many larger nebulæ. This is quite consistent with the idea that they consist of separate stars lying in a plane largely inclined to our axis of vision: but seems quite inconsistent with the idea that they consist of mere vapor manipulated by the laws of heat and gravity.

In view of such considerations, such and so

many, it seems little to say that the evidence is far from being decisive in favor of the so-called gaseous nebulæ being severally fire mists. Indeed, the evidence seems decisive against it. But granting that it is not — granting that, despite all these urgent appearances to the contrary, the nebulæ giving bright spectral lines have not been misinterpreted and are really so many huge banks of glowing vapor — what then? Does it follow that we have at last found the raw material out of which mere Nature can manufacture heavens, with all their innumerable flocks of globed and shining whirlwinds?

This brings us to the third proposition which we set out to examine, namely, Granting that some nebulæ are fire mists, it is plain that they are not such fire mists as the Nebular Hypothesis demands.

This hypothesis demands, not only fire mists, but fire mists of certain numbers, sizes, specimen stages, and chemical constitution.

As to *number*. Since all stars, groups, and clusters are supposed to come from so many fire mists and to return to the same, we ought to find as many of the latter as of the former. But we do not. Very far from it. The nearly six thousand nebulæ of all sorts known to us are quite inappreciable in the presence of the hosts of stars ; much more the nebulæ which retain an irresolvable aspect in the Rossian reflector (a very small part of those before set down as irresolvable) ; much more still those of the unresolved nebulæ which give the bright banded spectrum. This spectrum is found belonging to only about one third of the nebulæ actually examined, even when selected as the most likely to be gaseous. It is imputed in a way of induction — only a few of each class having been carefully questioned by the spectroscope — to the planetary, annular, and irregular nebulæ : but we have on our list only thirty-four planetary nebulæ, four annular not planetary ; and the irregular are hardly one tenth of the whole number. Who can suppose these few fire mists, these mere gleanings of the heavens, to fairly represent the cradles and graves of all the celestial nations as they come and go ? None who listen properly to the modern science of probabilities ; for, just here, that science throws the whole Multiplication Table at the Nebular Hypothesis. Doubtless, for every solid orb that is seen, we ought to be able to show a nebula sufficient to make it.

Then as to *size*. The size of the largest regular nebulæ claimed to be gaseous is not what it ought to be, considering that these represent the great stellar groups and clusters of the universe. They ought to be as much larger than these clusters as these when resolved into gas would be larger than they now are — that is, almost unspeakably larger. But in fact they are much smaller. Where are the shapely and mighty fire mists that might be the parents of such groups as the Pleiades? Astronomers have not yet found them.

Another point. Among the thousands of nebulæ claimed to be fire mists — not to say, among a number as great as that of ripe stars, groups, and clusters — we ought to find many examples of each of the *principal stages of world structure* supposed in the Nebular Hypothesis: among others, many examples of a nebulous ring about a central nucleus; many examples of such a ring parting into several unequal nuclei; many examples of such nuclei drawn around the largest as if about to be absorbed (small stockholders around some great one); many examples of several riper nuclei outside of a ring with a condensed center. Bringing our telescopes to bear on the sky, what do we actually find? Just nothing whatever that can pass for an example of most of these stages: and it is very doubtful whether we even find anything that can stand for an example of the simplest ring-stage. The annular nebulæ are often cited to the contrary. But only about half a score of these objects have been found in both hemispheres; and most of these few are mere rings with dark interiors, and so are not examples of such ring-nebulæ as the Nebular Hypothesis supposes. Of two exceptions, one ring is only slightly luminous within and has been quite resolved into stars: while the other, 51 Messier, with a dense center, is heavily split through the greater part of its circumference — a fact hardly consistent with the idea of a ring formed by rotation. Besides, only four of the annular nebulæ have as yet been shown to give the gaseous spectrum: further observation may show, as has been done in the case of the spiral nebulæ, that part give the continuous spectrum.

So the ring-stage of the fire mist is very slenderly, if at all, represented in the actual sky. And yet this is the leading stage in the Nebular Scheme. Indeed, it mixes itself up with nearly all the other stages. We ought to see the ring, more or less advanced, in most examples of gaseous nebulæ — not to say several rings at once. That we *never* see it, or next to never, is a thing to be complained of by the evolutionist. Not by me. I am very well satisfied with the heavens as they are. I wish that Rings, especially gaseous ones, were as scarce below as they are above.

But there is another point of still more striking character. The Nebular Hypothesis requires the gaseous nebulæ to be *composed of as great a variety of elements, in the same general proportion, as our sun and the stars.* But the solar spectrum is striped with more than two thousand lines — the principal stars, such as Vega, Capella, Aldebaran, Sirius, show spectra hardly less rich — some fifty other stars have been put to the question and have reluctantly confessed to similar wealth — in fine, there can be little doubt that nearly all the stars are made up of a great variety of elements. We ought to find a similar variety in the gaseous nebulæ, if they do indeed stand for the stars in their first stages. But, actually, we find all these nebulæ about as poor just here as they can well be without being quite bankrupt. None of them give more than four or five bright bands in the spectroscope, most of them give only three, and some give only one band. In general, only nitrogen and hydrogen and, perhaps, another unknown element, appear in the spectrum. If this tells the whole story, it is indeed a very sorry chemical treasury which the gaseous nebulæ have as compared with that of the flushed and wealthy stars. Hardly enough to keep them in countenance, hardly enough for daily bread, certainly not enough for the society they move in. But is this the whole? May there not be in the background other elements too weak to show in so faint a spectrum? The brightest gaseous nebula, that in the sword of Orion, gives only three bands: a much fainter spiral gives four. The Dumb Bell Nebula, and that in Aquarius, which give each but a single bar, are brighter than many a planetary nebula which gives a triplet of bars. Besides, in the stellar spectra generally there are many other lines quite as strongly marked as those of nitrogen and hydrogen. Often more strongly than these. Sometimes these two elements do not report themselves at all. This shows that if other elements exist in the gaseous nebulæ they ought to show themselves in the spectra: for, the supposition is that their elements are in general proportioned to each other as they are in the stars.

Hence, some three elements express substantially the whole contents of nearly every gaseous nebula. And these elements are not only the same in number for nearly every nebula, but they are also the same in kind and proportion; while the elements of the stars differ widely from each other in these respects. Also, nitrogen is invariably the leading element in the gaseous nebulæ, while far from being so in the stars. Of course such nebulæ cannot be the parents of such stars — the one class almost emptiness itself, and the other as swollen with chemical riches as ever were western clouds with the rainbows of departing day; the one class always dominated by nitrogen, and the other very far from showing any sign of such fealty; the one all alike in their elements, and the other as various as the devices of heraldry or the gay ribbons of commerce.

Thus we have gone over the three great astronomical fields — the solar system, the stellar systems, and the nebulæ. In each field we have been met by numerous facts in apparent conflict with the Nebular Hypothesis. We found our sun still glowing with a heat and light so enormous as to repel the idea that it is but the embers of an exhausted conflagration: we found the chemical constitution of our system so various in different parts as to discourage us from thinking that its members were all formed out of one boiling mass, as well compounded as ever were the anxiously stirred and shaken preparations of the apothecary or chemist: we found its members differing among themselves as to size, density, moons, atmospheres, water, in so irregular a way as cannot be reconciled with the idea that they were all formed from the same mist under the same general circumstances, or circumstances changing steadily according to a simple law: we found that both the rotations and the revolutions of the system fly in the face of a theory which requires them all to be easterly, all circular, all in the plane of the sun's equator, and all under the same laws generally for planets and for satellites.

Passing to the stellar systems, we found that many of them are of great visible size; are without dominant central orbs; show no graduation from the center outward in the light, distance, and size of its members; have their orbits very elliptical and in widely different planes; differ exceedingly in chemical constitution, whether we compare the stars of the same group or of different groups with each other — being in all these particulars just the opposite of what we are taught by the Nebular Hypothesis to expect.

Passing on still to the great nebular field, we found that, while the chief findings of the spectroscope are explainable on the supposition that all the nebulæ are clouds of separate stars, there are great positive difficulties in the way of supposing that they are merely vast oceans of thinnest vapor — in the facts that they are seen at such immense distances, that they often appear under very irregular shapes, that they maintain great permanency of general aspect both as to contour and interior configuration, that they often are sharply defined and uniformly bright throughout. Indeed, we found that if we grant that the nebulæ are mere fire mists, they cannot be accepted as being *such* fire mists as would be available for making bodies like the planets, sun, and stars — because too few, too small, too scanty in specimen stages, and especially too poor in chemical constitution.

These facts, you observe, are very many and various. They belong to the latest researches as well as to the most ancient. They are gathered from widely different fields as well as widely different periods — indeed from all the great astronomical fields accessible to us. And they are not merely facts which will have nothing to do with the Nebular Hypothesis, of which we have a plenty — not merely facts shooting at it a whole font of interrogation points, of which we have more than a plenty — but facts that directly assail it with aspect and gesture of battle. They assail a scheme already encumbered by a host of facts hanging at its skirts, and teasing for explanation of themselves with incessant clamors which can neither be satisfied nor silenced. In this hampered condition, what has it for defense, save such Swiss soldiers as can be hired to defend almost any hypothesis — that is to say, certain scattered agreements with Nature. An hypothesis must be terribly outrageous not to have *some* verisimilitudes about it. No error, no heresy, no Satan even, that has ever offered itself to the world in either art or science or religion, has been without some bits of the wardrobe of an angel of light. But, then, these defenders of the Nebular Hypothesis are very curious defenders after all. They fight just as hard for the Theistic Hypothesis as they do for the other. They agree at least as well with the idea that the worlds were created by eternal God as they do with the idea that they were developed from eternal matter. When pushed into the foreground by evolutionists, to do battle in their behalf, they flatly refuse to serve. They are neutrals. They sit on the fence. They neither help nor hinder — until it is plain which way the battle will go. Then they help the winning party. Just as soon as Nature begins to give way before the Supernatural they begin to discover what side they are on. "This is our side," they exclaim, "it has *always* been our side;" and away they dash in chase of the fugitives after the most approved manner of the chivalry. I shall not condemn them for this. It is altogether proper — as well as exceedingly human. Who does not know that when faith is once victorious every verisimilitude about it adds to its strength? Really — so lame, so embarrassed, so poorly supported by its nominal friends — the Nebular Hypothesis is not well defended against the three formidable archers who have agreed to play against it from the hights of their azure fortress. Three archers; Robin Hood, William Tell, Aster of Amphipolis — three silver quivers; each full of arrows, tough, sharp, well-feathered, addressed to Philip's right eye — three sounding bows; borrowed from Diana and Phœbus and all the heavenly host; bent by three athlete Astronomies, till the tips meet; bent against the latest atheism, because against the only scheme that in these days tries to explain the heavens without a God!

The case seems to be this. A man sits down in his study and draws a chart of certain seas. He has never been on them. No voyagers have ever described them to him. Much less has some Coast Survey Commission carefully sounded and triangulated for him the whole district. But he has a certain geological idea: or he has a certain idea mythological; and has been told that the giant Typhon, so many miles long, was once cast down into that district and outlined its surface by his prostrate and buried form: or he has stumbled on the notion that the irregularities of the earth's surface are a sort of mineral vegetation, say a mineral cucumber, whose law of stock and branch and leaf is well known: or he has made the discovery that the earth itself is but a huge animal, whose whole frame and exterior, as of some Mastodon, can be scientifically divined from any bit of it however small. So he sets to work. He maps down the district as it ought to be on his principles. That the map may be attractive as well as true, he paints it. Then he passes it over to a sailor for a practical trial. Up with the anchor, fling free the sails, grasp the wheel with your hands and the chart with your eyes — now see what we shall come to! Coasting along, we notice here and there a headland which seems to answer tolerably well to the chart; but, meanwhile, we are constantly finding, on the right hand and on the left, objects which do not appear on the paper at all. But this is not the worst. Sometimes, where the map says fifty fathoms, we sound and find five — sometimes, where we are bid to expect good anchorage, we cast out and find the worst bottom possible — sometimes, where a strong current is set down with its heavy arrows, we find still water; or, where the flow is set down as being easterly, we find it westerly — sometimes, we come across a strait where there should be a sound; a shoal, a reef, an island, a continent where should be unobstructed sailing. How long would we go on in this way before giving up our remarkable chart and the principle on which it was made? Not more than a thousand miles. The occasional agreements would go for nothing. The numerous disagreements would quickly wreck our confidence, if they did not our ship. We fling the worthless paper away with both hands — though on one corner of it may be seen, "G. HERSCHELIUS PINXIT."

So a man makes out, on the principles of the Nebular Hypothesis, a map of what we ought to find in the skies. I take it on trial, and set out to navigate with it celestial seas. "Ho, men! Shake out every rag of canvas! The way is long and we have no time to lose." Away goes our adventurous Argo before the wind; and, sure enough, it is not long before we are obliged to confess to finding, on the celestial coasts, a bluff here and a turn in the channel there that very well agree with the description. But then, where are *these* features of the azure oceans, and *these*, and *these* — why do they not appear on the map? "But sail away, O Jason, and see if the sailing does not improve! You have only just begun your voyage. Perhaps you will yet find the Golden Fleece." So we go on — go on to bring up squarely against astronomies, which, according to the Nebular chart, not only ought not to be there, but whose place ought to be held by their direct opposites — brightness where should be dimness; diversity where should be sameness; irregular sequences where should be regular; westerly motions where should be easterly; individuals where should be groups; ellipses, and sometimes almost parabolas, where should be circles; discord of orbits, largely perpendicular discord, where should be exact and immovable concord ; vacancies where great worlds should be, and worlds where great vacancies should be ; nearness where should be remoteness, and remoteness where should be nearness ; largeness where should be smallness, and smallness where should be largeness ; many where should be few or none, and few or none where should be many — and so on to the fiftieth astronomical breaker. Pray, how many scores of times shall our chart bring us up against such forbidden and unexpected shoals, reefs, islands, continents, before we ought to see our way clear to conclude that the chart is worthless or worse, and that the curious principle on which it was made is utterly fallacious !

A word in your ear. It is a wonder that our audacious vessel has not been quite wrecked under such piloting. Let us venture no further. And let us join in throwing overboard the idea that some prostrate Typhon is outlining with his huge form the celestial geography — that the starry universe is only a sort of mammoth vegetable or animal whose whole can be described as soon as a single piece is found — or, what is about the same thing, that a chaos of smoke and bedlam of disjointed atoms can manufacture itself, not only into the orderly and august schemes of the firmament, but even into such elaborate organic and spiritual beings as crowd the surface of one at least of its innumerable orbs.

> "From west to east the earth
> Unrolls her primal curve;
> The sun himself were vexed
> Did she one furlong swerve:
> The myriad years have whirled her hither
> But tell not of the whence and whither."
>
> "We know but what we see—
> Like cause and like event;
> One constant force runs on,
> Transmuted but unspent:
> The natural choice that brought us hither
> Is silent on the whence and whither."
>
> "If God there be, or gods,
> Without our science lies;
> We cannot see or touch,
> Measure or analyze:
> The self-moved force that brought us hither
> Reveals no whence, and hints no whither."
>
> If *this* be all and all—
> Life but one mode of force;
> Law but the plan which binds
> The sequences in course:

All essence, all design
Shut out from mortal ken
We bow to Nature's fate,
And drop the style of men!
The summer dust the wind wafts hither
Is not more dead to whence and whither.

I sympathize with the scholar of Baliol in the mingled shame and wrath with which he wrote these lines. In the face of the heavens and earth — what a philosophy! Our common sense puts it away with both hands. Phenomena are *not* all that we know. We know also the Whence and the Whither. Tradition looks about on wonderful Nature, and then points *upward* with her finger of mist. Science looks about on a Nature still more wonderful, and then points *upward* with her finger of stone. Revelation looks about on a Nature — Oh, how much more wonderful still — and then points *upward* with both hands and with all her fingers of gold. Following with our eyes those significant fingers — up through transparency after transparency, through azure after azure, through vacant infinity after vacant infinity — we come at last, not to a brute fog and miserable jumble of know-nothing mechanics and chemistries that somehow manage to swing from everlasting to everlasting through paradises of order and beauty and construction, but to a sceptered PERSON whose glory abashes and rebukes all human words. That scepter waves, and from its diamond tip leap worlds, systems, universes. That scepter waves again, and straightway the naked worlds are clothed with more than the jeweled robes of Solomon. Waves the scepter still again, and at once the miracle animals take their places in the ready palaces of sea and air and land. Waves again, and still more emphatically, that scepter, and lo, souls, with their constellation-faculties and glorious orbits of thought and hope and achievement and virtue, leap forth in still superber astronomies to reign over all. Behold the WHENCE — the WHITHER also!

This is the higher philosophy. And yet it is the philosophy with which we started in life. From the dear lips of sainted fathers and mothers we long ago heard of Him "who spake and it was done, who commanded and it stood fast." And now that we have lived to lift for ourselves just a corner of the veil which screens the magnificence of Nature, we see no reason to go back on the teachings of our childhood, but rather reason to say that such wonders can only be creatures of law by being at first hand creatures of God.

Creatures of God let us call them — and so repeat the venerable traditions. Creatures of God let us call them — and so affirm anew the grandest and most useful fact the world has ever known. Creatures of God let us call them — and so put our science at one with the religion that has ever been saying, "We understand that the worlds were framed by the word of God, so that things which are seen were not made of things which do appear." Creatures of God let us call them — and so have a chart by which we can easily find our way in the darkest night and under clouds of swollen canvas, not only amid the shining Polynesias of the sky, but also amid the more difficult and more shining Micronesias of organism and spirit that so thickly spangle the floods and fields of our own world; and the smallest of whose glorious islands, whether constellated below or constellated above, is both a mystery and a breaker, save in the light of GOD.

omy, as a whole, from it than I ever got before from all other sources, — more than from Enfield's great book, which I once carefully worked out, eclipses and all.

"I trace the progress made, and the methods of the same, and seize on the exact status of things at the point now reached."

From the Bibliotheca Sacra.

"This is a remarkable book, — one of the most remarkable which has proceeded from the American press for a long time. It lifts the reader fairly into the heavens and unveils their glories. The presentation is very full though concentrated, very clear and animating, — with a command of language and a glow of eloquence which is quite extraordinary. The last lecture is hardly less than a Te Deum. The only adverse criticism which, on reading the preparatory lecture, we were inclined to make, was, that the almost impassioned eloquence with which it opened would have been more impressive further on, and after the imagination had been excited by the facts. But, after finishing the last Lecture, we could not wonder that a mind so full of the great facts, and of the emotion which they necessarily kindle, should, on seeing his own parish charge assembled to listen, break forth in strains which none but a mind fully roused by his theme and his audience would have been able to utter. No person can read through this volume without mental exaltation, and a conviction of the peculiar ability of the author."

From the New Englander.

"It presents an admirable *résumé* of the sublime teachings of Astronomy, as related to natural religion, — a series of brilliant pen-photographs of the Wonders of the Heavens, as part of God's glorious handiwork. The first five lectures pass the science in rapid review; the last treats of the Author of Nature, as related to its leading features. There is not a dry page in the volume, but much originality and vigor of style, and often the highest eloquence. It is, withal, evidently by an author at home in his subject, not "crammed" for the task. It affords a fine example of what an intelligent pastor can do, outside of his pulpit, towards training an intelligent people, and by imparting to them Nature's teachings, leading "through Nature up to Nature's God," — the God of Revelation as well. To such a book the author need not hesitate to affix his name."

From Rev. A. P. Peabody, D.D., LL.D., Preacher to Harvard University, and Plummer Professor of Christian Morals.

"Permit me to thank you for a work in which you have effected a rare union of scientific accuracy, eloquent diction, and rich devotional sentiment. It is attractive, instructive, and edifying. It appears at a time when science needs, as never before, to be redeemed and sanctified by faith in Him, in whom are hidden all the treasures of wisdom and knowledge. And, best of all, it does not make Religion cringe to Science, but maintains her in that queenly status which is the only position she can hold. The book must do great good, and I heartily congratulate you as its author."

From Rev. S. H. Hall, D.D.

"Ecce Cœlum is much more than a book-success. It will be honored as a most timely and admirable treatise to put into the hand of thoughtful young people, to 'turn off their minds from vanity,' and lead them to God."

From the New-York Evangelist.

"This unpretending, though elegant little volume, gives a most admirable popular summary of the results of Astronomical Science. The author has evidently mastered his subject, and he has presented it in a most striking manner, adapted to the comprehension of the common reader, and enriched with pertinent illustrations. The book is perhaps the most fascinating treatise on the science which has been published of late years, ranking indeed in many respects with that of the late lamented and eloquent Mitchell. One of its excellencies is that it does not hide God behind his own creation.'"

From the Religious Herald.

"A New Book, and one that is a book, worth its weight in gold or diamonds, for it is full of gold and precious gems, — diamonds of law and fact, — truths beaming with celestial light. I speak of 'Ecce Cœlum,' from the pen of Rev. ENOCH F. BURR, D.D., of Lyme, Conn., published by Nichols & Noyes, Boston, a duodecimo of 198 pages. Mr. Burr modestly signs himself 'A Connecticut Pastor,' but some college has rent the vail and written out his full name, and added to it a D.D. So much the better for Connecticut and for the world. Such light as the book contains ought not to be under a bushel.

"These six Parish Lectures are a masterly, vivid, easy, sublime presentation of the enchanting facts of Astronomy. They are adapted to all classes, — the learned and the unlearned. The astounding glories of the skies are tempered to our humble eyes.

"Let all read the book, old and young. Let it be found in every school, in every library, and in every home where wisdom is invoked. Read it, and you will exclaim, what glorious light it sheds from the throne of God upon the lonely pathway of man!"

From C. H. Balsbaugh, of Pennsylvania.

"It is certainly a wonderful little book. How the world shrinks into an atom as we follow the lofty soarings of the 'Connecticut Pastor.' I never knew rightly what Dr. Young means by saying, 'an undevout Astronomer is mad;' but I now see and feel the power and beauty of the expression. Such a book cannot be read without laying upon us the responsibility of a new charge from heaven. After contemplating such grandeur, we instinctively exclaim, 'What is man that Thou art mindful of him?'"

From Hon. S. L. Selden, Late Chief Justice of New York.

"A beautiful book. I admire it for the elegance of its style, as well as for the lucid and able manner in which it presents the noblest of the sciences. It will prove, I think, very valuable, not merely for the knowledge it communicates, but as suggestive of a line of noble and elevated thought. And I am much pleased to see from the numerous notices which have come under my observation that my estimate is confirmed by many persons of the first capacity for judging. To have written a work which receives and deserves such *very* high praise from scholars and men of science cannot but be a source of great gratification to the author."

ECCE CŒLUM;

PARISH ASTRONOMY.

ELEVENTH EDITION.

SUPPLEMENTARY EXTRACTS.

From the Theological Eclectic, [*Edited by Professor Day, Schaff, etc.*]

"The style is remarkably graphic and elastic, and the matter is so skilfully grouped and lucidly stated as to be level to all classes of readers. The writer has a rare gift at popularizing science, and his book deserves the wide welcome it has received."

From the New York Observer.

"We have never yet seen a volume on Astronomy that seemed to us to explain more intelligently, to ordinary minds, the visible phenomena of the heavenly bodies."

From the Congregationalist.

"We advise all our readers who have not yet read the book entitled 'Ecce Cœlum,' to embrace their earliest opportunity to do so,—a book which certainly has been surpassed by nothing of this general line, for many years, if ever. There is a grandeur of conception—an easy grasp of great facts—a clear apprehension of deep and subtle relations—a power to see, and make others see, the nature and extent of the heavenly movements, such as are altogether wonderful. Many works have been written from time to time to popularize astronomy—to bring its great leading features within the compass of unscientific minds. But we do not know of a work in which this has been so finely done as in 'Ecce Cœlum.' Six lectures of about an hour each, tell the story, and the reader feels, all the while, as if he were upon a triumphal march. He is upborne and sustained by his guide, so that he has no sense of labor and weariness on the journey. The last chapter, on 'The Author of Nature,' is a most worthy and fitting close to the book. We wish it could be read by that great host of so-called scientific men, who are delving away in the mines of nature, with thoughts and purposes materialistic and half atheistic. They need the tonic of such Christian thinking as this."

From Hours at Home.

"This little book, from the pen of Rev. E. F. Burr, D.D., has already been noticed extensively and pronounced a 'remarkable book' by our best critics. The author first delivered the substance of it to his own people in familiar lectures. It presents a clear and succinct resume of the sublime teachings of astronomy, especially as related to natural religion. The theme is an inspiring one, and the author is master of his subject, and handles it with rare tact, and succeeds as few men have ever done in giving an intelligent view of the wonders of astronomy, according to the latest researches and discoveries. It is indeed an eloquent and masterly production."

From Harper's Monthly.

"The title page of 'Ecce Cœlum' is the poorest page in the book. We have seen nothing since the days of Dr. Chalmer's Astronomical Discourses equal in their kind to these six simple lectures. By an imagination which is truly contagious the writer lifts us above the earth and causes us to wander for a time among the stars. The most abstruse truths he succeeds in translating into popular forms. Science is with him less a study than a poem, less a poem than a form of devotion. The writer who can convert the Calculus into a fairy story, as Dr. Burr has done, may fairly hope that no theme can thwart the solving power of his imagination. An enthusiast in science, he is also an earnest Christian at heart. He makes no attempt to reconcile science and religion, but writes as with a charming ignorance that any one had ever been so absurdly irrational as to imagine that they were ever at variance."

From the Evangelist.

"We have had many inquiries in regard to the authorship of Ecce Cœlum,' the volume noticed somewhat at length two weeks since. To save writing a number of letters, we may say here, that the Country Pastor, who is the author of these six Lectures on 'Parish Astronomy,' is the Rev. E. F. Burr, D.D., of Lyme, Ct. The book is a 16mo of about two hundred pages, but in that small compass it comprises the results of long study, and will be found as instructive as it is eloquent. The grandest truths are made level to the plainest understanding. We took it up, expecting little from its humble pretensions, but soon found that it was all compact with scientific knowledge, yet glowing with religious faith, and were not surprised that Dr Bushnell should say he 'had not been so fascinated by any book for a long time — *never* by a book on that subject' — and that it had given him 'a better idea of astronomy than he ever got before from all other sources.' We don't know if they have many such ministers 'lying around' in the country parishes of Connecticut, but if so it must be a remarkable State.

"While the impression of this fascinating volume is fresh in mind," etc.

From Rev. G. W. Andrews, D.D., President of Marietta College.

"The author has succeeded admirably in his attempt to present the great facts of Astronomical Science in such form as to be intelligible to those who have not gone through with a thorough mathematical training, and to make them intensely interesting to all classes of readers. I cannot express more strongly the interest the volume excited than by saying that I read through at once. I can hardly remember when I have done the same with another work."

From Rev. Edwin Hall, D.D., Professor in Auburn Theological Seminary

"I received it last night, and have read it through with intense interest and delight. It is a worthy book on a mighty theme. I wish it might be in every household, and read by everybody. And I am sure it will be read with admiration and wonder long after the author shall have been gathered to his fathers."

From Rev. Prof. E. W. Hooker, D. D.

"The book is an admirable argument from the discoveries of modern Astronomers, for the existence of God; and indirectly for the truth of the Gospel. It is an honor to his kindred, to the

Church and the place of his birth, and, above all, to Him whose gospel he preaches."

From an Obituary of Rev. S. L. Pomroy, D.D., late Secretary of the A. B. C. F. M.

"He was a man of extensive information, a ripe scholar, and he retained his scholarly habits and tastes to the last. A few weeks since he read 'Ecce Cœlum' with great pleasure and satisfaction, When he returned it he remarked, 'I have read it all twice, parts of it three times, and have noted down certain passages.' He was specially delighted with the arrangement of the work — the grouping of the different system so as to give us something like a comprehensive idea of the grand whole."

From the Congregational Quarterly.

That a Connecticut Pastor should be able in six lectures to his peeble to shed more light on this profound subject — to make it more simple and yet more grand, amazing, and impressive — than many of the great masters who have written before him is a matter of surprise. Yet this seems to be the generally conceded opinion of the press. We hear but one testimony concerning Ecce Cœlum. Any intelligent reader of it can understand what before has been only a mystery. It is worthy of the widest circulation.

From the Lawrence American.

There is not a dry page in these six lectures; but the glories of the skies are presented in a most enchanting manner, vivid, popular, grand, and glowing. Young and old should read it.

From The Christian Union.

We can commend this book in the heartiest manner. It is one of the noblest examples of the moral uses of astronomy that have appeared since CHALMER'S astronomical sermons. Besides their intrinsic merit, these lectures show what may be done by a quiet pastor of a village church for the instruction of his people. Every preacher has not the equipment required for a course of scientific lectures: but "where there is a will there is a way," and much more might be done than is done in broadening a pastor's literary education and in raising the literary tastes of his people.

EXTRACTS FROM NOTICES.

From the Rev. W. A. Stearns, D.D., L.L.D., President of Amherst College

I have heard them with the deepest interest. They are so clear, so logical, so rich in illustration, so unexceptionable and beautiful in style, and so conclusive in the argument attempted, that I have profoundly admired them. Those gentlemen who heard them when delivered here, would, I am sure, from the comments which they made upon them, agree with me entirely in the judgment I have expressed. May the Great Being whose existence these lectures so nobly defend from the attacks of the foolish, though calling themselves scientists and philosophers, spare the life of the author and enable him to complete the full course of thinking on which he has so triumphantly entered and advanced.

From Rev. Prof. C. S. Lyman, of Yale College.

All whom I have heard speak of these lectures have expressed for them the highest admiration. In thought and diction they are worthy of Chalmers.

From Prof. Julius H. Seelye, Professor of Mental and Moral Philosophy in Amherst College.

It is with great delight that I have received the new book. I like, especially, its whole attitude respecting the question discussed; that it is so full of faith and so uncompromising. Atheism is as unworthy the intellect, as it is repugnant to the heart; and I am tired of tame apologies from timid believers in a God. I like to see a book that has something of a clarion ring about it, and is not afraid to defy denial, when it speaks of the being and the glory of the Heavenly Father.

I believe that Pater Mundi will do great good, and I thank the Lord for permitting the author to prepare and publish it.

From Rev. A. P. Peabody, D.D. L.L.D., Preacher to Harvard University, and Plummer Professor of Christian Morals.

I thank the author with all my heart for Pater Mundi. It is the most efficient work of its class which the present generation has produced; and as the now existing scepticism is deeper, more [pseudo] scientific, more pretentious, than that of any preceding age; the book which, like Pater Mundi, is adapted to our times, must need be both broader and more profound than previous needs have elicited. Its treatment of the great theme is at once thoroughly philosophical and popular, both in style and in adaptation to the capacity of all readers of average intelligence. It was an unspeakable privilege to the students of Amherst College, to have heard the lectures; I trust that the same privilege will be extended through the press to thousands of our young men. While I find no fault nor deficiency in the treatment of any branch of the argument, I am especially impressed by the Seventh Lecture, as the clearest, strongest, and most eloquent statement of the need of God, and of the demonstration thence resulting of His existence, in the plenitude of His attributes, that has come within the range of my reading.

From Rev. Albert Barnes,

I was so profoundly impressed, or, if I may say so, *oppressed* and overwhelmed with the sublimity and grandeur of the truths presented in Ecce Cœlum, and with the manner in which the author presented these great truths, that I am glad he has followed with another volume on the same general subject. I anticipate in the perusal of it great pleasure and profit. I think the author is doing great service to the cause of truth and I hope that God will spare him to complete his work.

So far as I am able to judge, the greatest enemy which Christianity has to encounter now, is found in the oppositions of science, so-called. In fact, so far as I understand them, the aim and tendency of much of this science, are to blank Atheism; and I think a man can do no better service in this age, than to meet and counteract this tendency. I rejoice that God raises up men who are qualified to do it. I believe that the author of Ecce Cœlum is such a man. He has a noble work before him, and I hope he will be enabled to do it.

From the Independent.

We had not read *Ecce Cœlum,* and imagined that the enconiums which we had seen pronounced upon it must be too high wrought for sober truth. But now that we have read *Pater Mundi,* by the same author, we are ready to believe every word of praise to have been within bounds. The present volume is no dry, didactic treatise. It is warm, alive, eloquent. The author proves himself, in his freshness of thought and in the eloquence of his argument, inferior to no writer of the day. We find no slips in science, nor in his multiplied illustrations from ancient and modern literature. And we do find a grandeur of conception and a striking originality of conception, so audacious that scarcely any other writer we know of would have ventured upon it. We see no reason why our author's writings should not become classics in the language. Nothing can be more invigorating to the thoughtful reader.

From the Congregationalist.

We have read it with keen enjoyment, and are disposed to regard it as he most substantial and serviceable contribution to the natural theology of this generation, as it is the freshest and most popular. No better book none more entertaining, can be placed in the hands of inquisitive readers, especially bright minded young men and women. The author lays out his work with a singularly clear perception of the crepuscular skepticism which needs to be dissipated; and enters upon it with manly and gener ous fairness of statement, vigor of argument, and amplitude of apposite and convincing illustration. His style is in the main so admirable, that it may seem ungenerous to take exceptions. Probably the excess of ornamentation, the overfulness of illustration, the easy affluence of the most highly poetic diction, and the general gorgeousness of rhetoric will secure a hearing for the truth by persons whom it is desirable to influence, who might not be attracted by an ordinary book.

From the Hours at Home.

The decidedly oratorical style will serve to make the essays, incisive—eloquent, and eminently philosophical as we acknowledge them to be—all the more widely popular and useful.

From the Religious Herald.

Cogent argument is so lighted up with brilliant illustration, as to make interesting the profoundest thoughts.

From the Christian Union. Rev. H. W. Beecher.

The author, who, in *Ecce Cœlum*, established a reputation for that rare combination of excellencies—fervid rhetoric, scientific accuracy, and common sense—has produced another book designed to defend and illustrate the doctrine of Theism. It is like breathing mountain air to feel this man's earnestness; it is a true mental tonic. One sees instantly that he is able-souled, that he can push and climb without getting short of breath; and it is almost a foregone conclusion, after reading the first chapter, that one must either stride with him to *his* high conclusion, or part company before starting. This unequivocal earnestness and power display themselves at the outset; great heart is warmed up to begin with; so that one is almost inclined to distrust a leader who has so much the air of a partisan. The face set like a flint does not wait to be struck to emit its sparks, but glows with a fiery zeal which inflames everything it looks upon. Yet, no candid reader will say that Dr. Burr is dogmatic; he only plies error with weapons for which infidelity has claimed a patent right. No one who reads this first volume, will wish that the author had written less or otherwise than he has.

From the Advance.

The previous work entitled *Ecce Cœlum*, received the highest commendation from the most competent judges. The present volume will still further augment the reputation of the author as a thinker and writer. He puts the Atheistic hypothesis to severe and annihilating tests; fully meeting its objections and cavils. The arguments of this work are not only cogent, but are expressed in a lucid, glowing, and eloquent style; and the book entitles the writer to a position among our best religious authors.

From Rev. Edwin Hall, D.D., Professor in Auburn Theological Seminary

I have read the work with constantly increasing satisfaction and delight It is entirely worthy of the author of *Ecce Cœlum* and of its subject. So far as my reading extends—and I have long endeavored to read in that department whatever I could lay my hands on that promised to give me light—I regard it as the most original and valuable contribution to the subject, which the age has produced. I shall wait with longing for the second volume. In the meantime, I hope the work may have a circulation as extensive as its worth deserves. If it were left for me to fix that desert, there should not be a library or a family in the land without it.

From the Watchman and Reflector.

The thousands of readers of "Ecce Cœlum" have not got fairly over the feeling of astonishment and admiration which the perusal of that remarkable book brought to them, before another of equal merit from the same author is announced. "Pater Mundi," we are confident, will lessen nothing of the high character which Dr. Burr has won as an acute and accurate thinker, an accomplished scholar, a brilliant rhetorician, and a humble, childlike believer in God and His revelation. The purpose of the author is to defend and illustrate Theism and Christianity from the side of Modern Science. There is a wonderful candor in the entire process of argumentation. Nothing is assumed beyond what the eyes of man behold and his reason assents to. The conclusion, without being asserted, is irresistibly forced into one's own view, and wins acceptance from the thoughtful, reasonable soul. The eloquence of some of these passages respecting the fatherhood of God is overwhelming in effect. We earnestly commend the book to the careful study of our so-called scientific men who are trying hard to rule a personal God out of the universe. We wish, too, that every young man in the nation would read these pages. We are sure that nothing more fascinating in interest and really healthful and elevating in influence can be found among all the books of the day. The book is handsomely printed by Nichols & Noyes of this city.

From the Sunday School Times.

This volume is an eloquent and unanswerable protest against modern atheism in all its forms. "Modern science testifying to the Heavenly Father," is the author's secondary title, and it describes accurately the course and object of his argument. His methods of presenting the subject, however, are entirely original, and are wonderfully effective. The work is particularly opportune. There are in all our congregations thoughtful, cultivated, quiet men, whose faith has been shaken by the bold assumptions of infidel scientists. Dr. Burr's book is just suited to restore such persons to their equilibrium. It is written in a most attractive style and shows a masculine vigor of thought that cannot fail to command respect.

From the Theological Eclectic. Professors Day, Schaff, etc.

We have already spoken of the able work entitled Ecce Cœlum, in terms of high commendation. The present work by the same author exhibits the same power of comprehensive grouping and vivid presentation, and abounds in great thoughts freshly put.

From Rev. Mark Hopkins, D.D.. L.L.D., President of Williams College.

I am greatly indebted to the author of Pater Mundi. It is a fresh and powerful work. If any commendation from me will aid its circulation, it is freely given.

From C. H. Balsbaugh, Pa.

Certainly this is a book to stop the mouth of skeptics. It seems to me that never was atheism in its protean forms more squarely met on its own ground, and never more clearly discomfited with its own weapons. No two links of its argument are left together. The author has triumphantly vindicated the title of his book. Its matter and style appeal to both our innate susceptibility to truth, and our sense of the beautiful. In my view, never did logic and poetry more heartily embrace each other; never did beauty smile more divinely on the face of the sternest facts.

From the New York Evening Post.

The clear and beautiful logic, and the crystal style of Ecce Cœlum, fascinated religious minds everywhere in this country. This book is written by the same perspicuous pen. That it is in the form of lectures, rather improves it than otherwise. The special aim of the author is to wrest from the wild materials of this day the powerful sceptre of science, which they have seemed to wield. All the teachings of science and nature point to the "Father of the World." This book is one calculated to strengthen the faith of professors of religion, and to lead captive young minds straying into error. We ought to mention in closing, the beautiful typography of the book. Published by Nichols & Noyes.

From the Evening Wisconsin, Milwaukee.

The style is clear, and always strong and forcible in an unusual degree while many passages rise to great beauty and eloquence. Seldom have we read anything upon the subject of Christian evidence that was so entertaining, so instructive, and so satisfactory as this book. It is the offspring, of a vigorous intellect, and it is a most valuable addition to religious culture.

From the Christian Recorder, Philadelphia.

So charmed are we with this magnificent production of Dr. Burr's, that really we scarce know where to begin its praise. Its excellence is uniform.

Lecture first and lecture eighth equally demand admiration. So every part of each lecture. The chain of gold is not only complete, but every link is complete. The Colonnade is not only symmetrical, but its minute carvings are perfect. To quote from it to our own satisfaction, would be to quote the whole book, but we remember that Messrs. Nichols & Noyes, the publishers, have a copyright.

How majestically does the author of Ecce Cœlum send forth his thoughts into the world! In majesty do they stride forth either to conquer, to convince, or to woo. Now as a mailed warrior are they seen, fully panopled from head to foot, and crushing by the strength of his arguments every foe—crushing every atheistic shield, and helmet, and breastplate. On almost every page of Pater Mundi, these all-crushing arguments are to be met—on almost every page we see victims lying mangled and bleeding.

We do not know that the author of Pater Mundi lays claim to the poetic gift; and yet has he given us a sublime Didactic Poem. Not in verse, is it given; it is neither Dactylic, Anapæstic, Iambic, nor Trochaic. But poetic imagination shines on every page. Untrammeled by rule, and enjoying a freedom that the utmost poetic license could not allow, the author has given us a poem infinitely sublimer than could possibly have been done in any other form. Would that we could give our readers the concluding pages of Lecture VII. Such poetic thought! Such beauty of expression! Such smoothness! Such harmony! Words answer to words, and sentence to sentence, with such sweetness that one glides along to the conclusion, as smoothly as a New England sleigh, and as merrily as its ringing bells.

From the Norwich Bulletin,

It will be a great advantage to the reader of this work to have made the acquaintance of Dr. Burr's previous volume, "Ecce Cœlum," as thus many of the references in "Pater Mundi" will be the more intelligible and vivid. The quality of the new work is in all respects admirable. Dr. Burr has a wonderful enthusiasm, always fresh and intense. He is full of his subject. He has the faculty of so treating profound and sublime themes, as to bring them easily to the comprehension of all. He has a fervid style, whose richness seems inexhaustible. He has great fertility in argument, and presents his suggestions with rare simplicity and force. The volume will go far to combat the sophistries of Atheism, both in uncultured minds and in those of strong logical powers. We cannot too highly commend it, and we predict that it will find a place in every well stocked religious library.

From the Standard, Chicago, Ill.

If any one should infer from the title of this book that it is a heavy and prosy dissertation, he would be astonished on looking over its pages

Nothing could be further from the truth. The author is an enthusiast, one of those who have not "discovered that one must be indifferent in order to be fair." The book is fresh, earnest, and eloquent, and we felt its strong spell before reading a dozen pages. The statement of arguments is admirably clear, the development of them is natural and impressive, and there is displayed a wonderful power in massing facts so as to give their full and combined effect.

From the Chicago Tribune.

This work in some respects is very remarkable. It is not only compact in argument, and forcible and clear in statement, but it is also absolutely brilliant and sparkling in manner, and rich and copious in illustration. Judging only from the one volume before us, we should pronounce it as one of the most remarkable and fascinating books of the day.

From the Orleans Republican, Albion, N. Y.

The author's premises are bold, and his line of argument clear, forcible and persuasive; shirking nothing, anticipating, and answering objections with equal fairness. The work is calm, liberal, and large thoughted; full of admirable logic, and profound reasoning; and the last three lectures, especially, are grand with beautiful and terrible imagery, exquisite poetry, and striking allusions to those mysterious facts and forces of nature which startle and awe believer and unbeliever alike; and his conclusion is singularly suggestive and powerful.

From Rev. Austin Phelps, D.D., Professor in Andover Theological Seminary.

I wish to thank the author for "Pater Mundi." Not that it needs any commendation from me: but I cannot but be grateful to any man who helps me to a new depth or vividness of conception of God; and this you have done by your book. I am specially impressed by the power with which it draws the great alternative, — a God benevolent, or a God malignant. The *reductio ad absurdum* is fearfully overwhelming; and the recoil with which one springs back from it gives one a lodgment and a resting-place in the Infinite Love which no gentler discipline could secure so well. This *vigor* of religious sensibility in your works charms me. We need it greatly in our Christian literature, to supplement alike the wiry intellect of which we have enough, and the emotive softness of which, perhaps, we have a little more.

From the American Baptist.

The author has a strong and vigorous style, and a power of grasping and grouping great truths, which make all that he utters luminous and convincing. Though prepared specially for educated men, they are adapted to all readers, have no abstruseness of diction, no intricate, far-fetched or dubious arguments. The author will impart no small measure of the indignation he feels towards atheism, concealing itself under the name of science, to those who read his book, and we trust it may have a very wide circulation.

From The New Englander.

The author of Ecce Cœlum could not well be expected to write a dull book on any subject, much less one in which God and nature were the chief topic. But whether he would be able to clothe a skeleton of a two-volume argument for Theism — often so dry and grim in other hands — with the flesh and muscle, the life and beauty, that charm us in Parish Astronomy, could only be shown conclusively by the production of a work like that before us. Pater Mundi, will, by the glow and magnetism of its rhetoric, and the enthusiastic earnestness of its tone, as well as the strength of its argument, be sure to command everywhere, appreciative and admiring readers, and prove, we trust, of special value to those who are inclined to regard science as hostile to religion. Its logic is vitalized and made effective by the force and richness of the illustrations drawn from the various fields of science. It is these all glowing often with poetic fervor, that rivet the attention at once, and carry the reader on insensibly from topic to topic. In some of the lectures, indeed, the argument assumes the elevation and almost the form of a grand poem. The sixth, for example, like a sublime ode, returns, strophe by strophe, with each point made in the argument, to the same exultant chorus, which becomes at once a ***quod erat demonstrandum*** to the understanding, and an inspiration of faith to the heart.

The second volume promises to be even more attractive than the first; for it is to be still more replete with the marvels and sublimities of the sciences as illustrative of the argument. It is too much forgotten by many that God may be studied in flower and forest, in storm and star, and in the soul of man, as well as in Moses and the prophets. The glowing pages of "Pater Mundi," teach impressively that the God of Revelation is the God of Nature as well.

From the Methodist.

The many gratified readers of "Ecce Cœlum." will welcome this new and important work of Dr. Burr. It is a book for the times. Natural Theology can no longer retain its old form: the progress, not only of Science but of speculative thought, demands a thorough revision, "Pater Mundi" meets this demand with masterly ability.

From the American Presbyterian Review.

A new work by the author of "Ecce Cœlum" is sure to attract unusual attention; nor will expectation be disappointed. Dr. Burr is an original and independent thinker, and he writes in a style of singular freshness and rhetorical beauty. His book is timely. Though popular in its address, it sacrifices nothing to effect, and is wholly free from that superficialty which is usually found in the attempt to reduce the conclusions of science to the level of a popular audiance. It discusses with masterly ability the testimonies of Modern Science to the being of a God, and defends Theism from the attacks of skeptical science in a bold and critical spirit worthy of all praise. It is as profoundly religious as it is thoroughly scientific. While it freely accepts the results of the freest investigations, it ably argues that there is nothing in one of these to shake the christian's faith, but much to confirm it. The work cannot fail to have an important influence on Natural Theology—bringing it into harmony with the progress of Science and speculative philosophy, and arming it with a new power of demonstration.

From the Princeton Review.

Dr. Burr, known to us in his youth as a modest and studious lad, and since, as the faithful and unpretending pastor of a rural congregation, has suddenly burst on our vision as an author of the first mark in the highest realms of thought, and as a leading defender of precious truth against the assaults of scientific pretenders and pretentious sciolists. He calls to mind the days when the great New England divines, the Edwardses, Bellamy, Backus, Smalley, Emmons, were pastors of agricultural congregations.

The universal approbation of Pater Mundi and the previous volume, by the press and by christian thinkers of the highest reputation, we find borne out by actual inspection. Real science is proved to be the handmaid of true religion, in a series of discussions which evince a masterly comprehension of the issues involved—a thorough acquaintance with modern science and its relations to religion—the whole in a style clear and simple, vivid and graphic. We think the quiet of a country charge more propitious to thorough study and deep thinking, than the din and whirl of metropolitan excitements.

From Prof. D. C. Gilman, Yale College.

I feel moved to express my hearty appreciation of the service the author of "Pater Mundi" is rendering to the world by the publication of these earnest, brilliant and impressive discourses.

From Hon. Henry L. Dawes, M. C.

The pleasure with which I read aloud to my family "Ecce Cœlum" has prepared me for an increased delight and profit in reading "Pater Mundi." I am very proud of the author, and rejoice in his growing fame.

From Our Monthly, Cincinnati, Ohio.

We are very glad to welcome and commend this book. The author does, with singular ability, what he proposes to do. His trumpet utters no uncertain sound. There is no danger of any one mistaking his meaning. We think it high time the arrogant assumptions and speculations of some scientific men in the interest of infidelity and atheism were exposed, and the harmony of all true science and revelation vindicated, made more apparent, and presented in some popular form. This Dr. Burr is doing, and the first installment of his work we have in this series of lectures. That they will be found interesting and convincing we need not say to those who have read "Ecce Cœlum."

From Rev. Roswell D. Hitchcock, D. D., Professor in Union Theological Seminary.

Its bright, fresh, vigorous rhetoric, is one of the least of its merits. Evidently the author has himself felt, and so has justly measured, the "oppositions cf science" which he combats. Only so can we get the confidence of thinking men, who are in trouble about the Bible. He does well to make so much of the moral temper of the inquirer. I often think that the apologetic literature of the Church, from first to last, has done little more than confirm and comfort those who were on the right side, and wished to remain there.

From Professor Taylor Lewis, LL. D., Professor in Union College.

I regard it as a very valuable contribution to our religious literature, and well worthy of the commendation the other works of the author have received. It is cheering to find that the many attacks on Christianity, under the names of science and free religion, are calling out so many books of intrinsic excellence. The great clamor of the enemy sometimes causes me to feel depressed; but such works as "Ad Fidem" assure me that there is power in the Church, both spiritual and intellectual.

From Rev. Austin Phelps, D. D., Professor in Andover Theological Seminary.

"Ad Fidem" has given me great satisfaction. It has been a greatly needed volume for a long while. What else have we in our literature on the Evidences which puts sound logic into readable style, so as to command the popular interest? I know of scarcely anything. Pastors are hard pressed, if I may judge from letters of inquiry which sometimes come to me, to find something which their inquiring young people will read by the side of the fascinating "Seers" of the Concord school. The author of "Ad Fidem" will find many to thank him for supplying the want.

From Hon. Jared B. Arbuthnot, LL. D.

Those who have known the author as one of the ablest mathematicians of the country; as a close student for years, and, almost to the sacrifice of life, of the profoundest branches of science; as a contributor to scientific journals of papers bristling with the utmost resources of the Calculus; and, latterly, as the author of a book on Astronomy, which has gone into many countries, drawn unprecedented eulogy from first scholars, and done more to make the most difficult of sciences intelligible and impressive to the general public than any other work ever written, will not expect to find him treating any subject superficially. They will not find him treating the Evidences in this manner. No reader of "Ad Fidem," who is himself a thorough scholar, will fail to see on every page of this, as well as of its companion volumes, under a popular dress, the order, thoroughness, immense force, and severe accuracy, as to both thought and expression, of a master in the exact sciences.

From Rev. A. P. Peabody, D. D., LL. D., Professor in Harvard University.

The author, or rather his numerous readers, should be congratulated on his continued and signal success in meeting the obtrusive skepticism of our times. His "Ad Fidem," in the choice and arrangement of topics, in its adaptation to existing needs, in soundness of reasoning, and in a vivacity and fervor which must command unwearied attention and interest, is precisely the work which the cause of truth demands. I am heartily thankful to him in behalf of the public for his service in the Gospel.

From Rev. W. S. Tyler, D. D., LL. D., Professor in Amherst College.

Clear as the air, bright as the sunshine, refreshing and invigorating as the northern breezes of this rare and beautiful season. There is in it a happy union of sound sense, good learning, personal experience, strong faith, and glowing eloquence, which bears the reader along as with an irresistible current. I admire particularly its boldness and directness. While there is sufficient moderation and prudence in stating the claims of the religion of the Bible, and the arguments by which it is supported, there is very little of the apologetic tone — there is no hesitation in appealing to the conscience and common sense of the *unbeliever* himself as on the side of the Christian Revelation.

I rejoice that the author has been permitted and enabled to add "Ad Fidem" to "Ecce Cœlum" and "Pater Mundi," and thus to lengthen and strengthen the chain which will, I trust, bind many to the truth.

From Rev. T. L. Cuyler, D. D., in the New York Evangelist.

Last evening my congregation enjoyed the intellectual treat of a brilliant discourse, by the author of "Ecce Cœlum" — that newly discovered star in our firmament of letters, in regard to whom so much interest is now felt. He is kinsman of President Burr, of Princeton College, and has devoted years to scientific studies. While listening to him, it seemed as if the frail form of flesh was ready to vanish away, while the inner soul was all aglow with the intense blaze of enthusiasm for the truth as it is in Jesus. His theme was — "The accord between the best literature and learning and the Word of God." It was a sparkling chapter from his newly published volume "Ad Fidem." The book abounds in sentences which are finished with the point of a diamond. Those who have read "Ecce Cœlum" will be hungry for this latest production of devout genius. The skeptic who can read its honest pages and not find his infidelity shaken, would hardly believe "though one rose from the dead."

From the Rt. Rev. Charles P. M'Ilvaine, D. D., D. C. L., Bishop of Ohio.

His admirable "Ecce Cœlum" had prepared the way in my house for its fit successor "Ad Fidem." In the range of its argument and in the force of its reasoning, added to the beauty and eloquence of its style, it is calculated to be, under the Lord's grace, eminently useful. The author appeals to evidences which none of the wisdom of the wise (of this world) can shake.

From the Springfield Republican.

"Ad Fidem" has met with much success — the first edition of fifteen hundred copies being exhausted within four days after publication. It is a vigorous and fascinating discussion of the Evidences of Christianity.

From the Interior.

The previous works of this author have been widely read, and much and justly admired. The volume before us is characterized by the same clearness and raciness, and will be read with interest by all classes.

From the Congregational Quarterly.

Dr. Burr has varied learning and remarkable rhetorical power. The earnestness and vigor of his faith are refreshing, particularly in an atmosphere surcharged with a speculative and skeptical spirit. "Ad Fidem" is well suited to relieve the doubts of the honest inquirer, and to strengthen the faith of the believer.

From the Literary World.

The author's fervor is exceedingly animating; the most indifferent reader cannot dwell unmoved upon his vigorous and glowing words; and those who reject his doctrines, must yield unqualified admiration to the skill and grace with which they are put forth. We have rarely fallen upon a professedly theological composition so rich in the genuine charms of rhetoric, so fascinating and persuasive in the delicate, yet forcible manipulation of grave and sombre subjects. Here is no dry discussion, no slow-going logical processes to disgust the reader with theme and thesis; the discussion is lively, the reasoning pleases while it convinces, and the impassioned earnestness of the writer allures his readers into willing tutelage, and brightens and beautifies his whole work.

"Ad Fidem" seems to us altogether admirable. It will bear and repay careful reading, for there has been no sacrifice of force to ornament. As a presentment of the claims of the Biblical religion, in a form at once universally intelligible and universally attractive, we know of no work which surpasses "Ad Fidem."

From the New York Observer.

"Ad Fidem" will, we believe, be greatly useful. It is admirably adapted to subserve the purpose designed. The author has made his mark as one of the ablest orthodox writers of the present day. He is a man of thought and study, and great power of expression. A short time since he burst on the religious mind of this country with a work called "Ecce Cœlum." He next appeared with a volume entitled "Pater Mundi," a profound, able, and timely series of chapters, proving that science testifies to the existence and attributes of the Christian's God. Modern professors of pure science would fain intimate to the world that it is unscientific to believe. Dr. Burr has made a book for these scientists and those who have been deluded by them to study. It is easy reading, and we recommend it to the learned and unlearned alike. It will do them all good.

From the Christian at Work.

It is a worthy compeer of his two previous volumes. Rhetorically, it is most brilliant. It is full of passages which break upon the soul like a revelation, and in following the line of his arguments, the reader cannot fail to be convinced that of a truth the Bible *is* God's holy Word.

We welcome it as a most efficient helper in setting at naught the efforts which are being made to cast contempt upon the sacred writings.

From the Boston Journal.

Another valuable addition to the solid and beneficial literature of the day, from the pen of the well known author of "Ecce Cœlum," and the almost equally admirable "Pater Mundi." The present work is a most excellent one, calculated in every respect to accomplish great and lasting good. The

Evidences of the Christian Religion are brought within the scope of average intelligences. The book fills a most important place in the domain of modern religious literature. The style is graphic, powerful, and elegant; and yet beautifully simple. His arguments, though conclusive, are within the reach of the unlearned as well as the accomplished. Nothing hard or pedantic characterizes any one of the sixteen essays of which "Ad Fidem" is composed; but the book is pleasant and profitable reading for everybody.

From the Methodist.

Dr. Burr's previous volumes have rendered everything from his pen welcome to thoughtful readers. His new book consists of *real* parish lectures. It is a book of evidences skillfully wrought out, and the better for being popular. The author always presents a happy combination of scientific information with cogent logic and a vigorous style.

From the Religious Herald.

We welcome another volume from the vigorous and attractive pen of the author of "Ecce Cœlum." For weight of thought, brilliancy of imagination, and force of style, it will compare favorably with his former works; and this is enough to insure for it an extensive sale.

From the Home Journal.

This book will doubtless attract more general attention and be more widely read than any previous work from his pen. The writer's scientific habit of mind and familiarity with the whole field of argument have enabled him to give the proofs of revealed religion in a clear and forcible style, in a way to aid many who are seeking settled religious convictions.

From the Watchman and Reflector.

The author who, a year or two since, so greatly startled the reading public by vaulting into a first place among Christian apologists, is likely to hold what he so splendidly won. This last book is, like the others which preceded it, in the interest of the Christian Faith. The pages sparkle with life. Its poetic fervor, its wonderful massing of facts, its brilliancy of illustration, its personal appeals, its resistless conclusions, make up a book which will not allow the most prejudiced or indifferent reader to lay it aside, when once it is fairly begun, until the last page is turned. It is the most successful attempt which has yet been made at popularizing the Evidences of the Christian Faith.

From the Western World.

The work is spoken well and widely of as a strong defense of Christianity against the growing materialism of the age. Its author has a high reputation as one of the most powerful orthodox writers of the country.

From the Evangelist.

It presents the various branches of evidence in a very eloquent and effective manner. Moreover, it is peculiarly appropriate to the present state of the religious world — establishing the foundations of faith in the Word of God, and vindicating the supernatural character of the Gospel of Christ.

From the Scottish American Journal.

The author of "Ad Fidem" is already famous to the world by his admir able little book, "Ecce Cœlum." His books are probably more highly and universally extolled than those of any other author — not excepting the author of "Ecce Homo" himself. "Ad Fidem" will undoubtedly add to Dr. Burr's fame. It is a popular religious writing of the highest order, that can be read by the masses, and that will not fail to accomplish a good mission. This book of itself is calculated to turn the tide against infidelity in favor of the good old-fashioned belief in the Scripture as the Word of God.

From the Utica Observer.

Dr. Burr's "Ecce Cœlum" and "Pater Mundi" have placed him among the foremost of modern contributors to religious literature. As a Christian writer, his characteristics are great clearness, boldness, and enthusiasm. He seizes the sword of argument, and gives no quarter to limping skepticism that quibbles over the Bible as a book whose Divine origin is undemonstrable. His arguments are presented with remarkable vigor and they cannot fail to strengthen the faith of the weak, and to "confound the foolish," who accept as confirmed a thousand facts upon far less evidence than we have of the truth of the Bible as the very Word of God. It would be difficult to find a volume of three hundred and fifty pages of recent publication in which is combined more of sound logic and religious fervor, or which is likely to result in greater good than this. Dr. Burr is a man for the weak Christian to lean upon; for the strong and confident one to esteem and admire, if not indeed to reverence.

From the Commercial Advertiser.

This is a very welcome book from the pen of the distinguished author of 'Ecce Cœlum" and "Pater Mundi." It is written at just the right time — at the time when the young men of the country show an unwillingness to "endure sound doctrine." Dr. Burr is a bold champion of the divine origin of revealed truth, and he handles skepticism without gloves. Let those who desire to know the truth read such a book as this. We do not fear the attacks of "scientists" upon revelation if those who read the speculations of science will, at the same time, exert themselves to reconcile history with Scripture prediction.

From the Advance.

A quite unanimous approval has greeted Dr. Burr in his labors as an author, as regards the value of his thoughts and the attraction of his style. The present work will meet with favor from those who appreciate the wants of our time. It aims to present the evidence in favor of the Bible, not in a dry, professional way, nor in a hot, polemic spirit, but with force and freshness, with appreciation of doubts and difficulties, and with the confidence of strong conviction. The author has much tact in coming at his subject, and his arguments are ingeniously constructed, and skillfully marshaled. He keeps in view, also, a practical result, and aims to impress the conscience as well as to enlighten the mind, insisting ever that the most solemn responsibility attaches to treatment of this great subject. We like the book, and wish it a large circulation.

From the Syracuse Journal.

Dr. Burr, the author, is a man in the prime of life, a lecturer in Amherst College, a man of profound scientific learning, patient study, and withal an earnest pastor, whose soul is aglow with enthusiam for the truth as it is in Jesus. His previous works, "Ecce Cœlum" and "Pater Mundi," have created a new sentiment in regard to religious subjects, and won for their author unbounded praise. They are notable books for the times, warm, alive, eloquent. "Ad Fidem" follows the path they marked out. In the words of Rev. Dr. Cuyler, "The skeptic who can read its honest pages and not find his infidelity shaken, would hardly believe 'though one rose from the dead.'"

From the North American Gazette.

The line of late publications indicated by "Ecce Cœlum," "Ecce Homo," etc., the first of which is from the same pen that now gives "Ad Fidem" to the world, can all be traced to the recent disputations in Europe over religious fundamentals. Of "Ecce Cœlum" we can hardly speak too highly to express the views of those concurring in its doctrine. It is thoroughly orthodox, compact, and thoughtful, and is a scientific as well as a religious essay; a work not unworthy to class with the great efforts of Chalmers. In half a dozen lectures it formulates more of the philosophy of orthodox faith than can be found in a century of ordinary sermonizing. This is the concurrent testimony of those whose opinions cannot be gainsayed.

"Ad Fidem" consists of a series of parish lectures, intended to settle the argument in behalf of the Bible. Of the execution of the labor too much can hardly be said. There is such an amount of plastic learning, close logic, and happy illustration, as justifies comparison with the astronomical discourses of Chalmers. Even the renown of Jonathan Edwards, so immovably crowned, is brought to mind by the closeness of the scientific analysis and synthesis used. And yet the whole is lucent to any ordinary understanding. The work takes instant rank with the foremost theological contributions of the day, and must exercise great influence.

From the Christian Recorder.

To secure the ready reading of "Ad Fidem" by those who have been fortunate to read "Pater Mundi," it is only necessary to inform them that it is from the pen of the same charming writer. It is a handsome book, and can be read with the most sensible joy.

It ought to be a question with thoughtful men, how these books of Dr. Burr can be placed in the hands of the people. We have not read "Ecce Cœlum," and consequently cannot speak personally of its worth. The others, however, we know to be books which the times demand. Could not cheap editions be issued — so cheap indeed, that the very widest circulation could be attained? With these in the hands of the class that make up the bone and sinew of the country, a strong bulwark would be erected against the rationalism of our German fellow-citizens, the papacy of our Irish, the infidelity of what few French we have, and the dizzy-headed nonsense of the few native-born Americans, who, to get notoriety, are willing to play the fool, in regard to the most vital of all subjects, religion.

From the Philadelphia Enquirer.

This volume consists of a series of lectures on the Evidences of the Biblical religion, delivered by Dr. Burr, the author of "Ecce Cœlum," a book which has gained a wide celebrity, in his parish in Connecticut. They were not originally intended for publication, but the author says that even if they had been they would hardly have been more careful in their statement of main facts and arguments. We do not think they would or indeed that they could have been much more exact or telling than they are. Dr. Burr is an advanced thinker, and a man of great liberality, so far as his books photograph him. His arguments are both cogent and persuasive, while through them breathes the all-powerful spirit of earnest conviction.

From the Congregationalist.

Some books are like a leaden rifle-ball; others like a cartridge, containing not only the ball but abundant means for propelling it. Dr. Burr's books are of the latter kind. This, his last, is not only a sound and good work, but it is active and stimulating. . . . We have a very able opening chapter entitled "Presumptions," which is worthy of being a book by itself so forcibly does it outline the grand general features of Christianity. . . . . Those who have read "Ecce Cœlum" and "Pater Mundi," will know what style to expect in the present volume. . . . . We accept this book as one of real power.

From the Lutheran Observer.

The readers of "Ecce Cœlum" and "Pater Mundi" — and their name is legion — will hail with delight this new work by the same "Connecticut pastor," who has so strikingly made the heavens declare the glory of God, and made the wondrous achievements of science testify to his wisdom, his greatness, his divinity and eternal power. It addresses itself to doubters and unbelievers with such an array of facts, and with such direct force of logic and argument, that it seems impossible for a rational soul to resist its conclusions. The book might appropriately be called rational and moral geometry, for its conclusions are the result of demonstrations as clear as any in Euclid.

The entire work characterized by great clearness and accuracy of style and statement, and it meets the objections and cavils of cultivated modern skepticism — the vague insinuations and sneers which float like froth upon the current of modern literature — better than other work that has yet appeared.

From the Christian Weekly.

"Ad Fidem" is a series of pastoral lectures to which the pastor has invited the reading public. And the reading public will be very apt to come when they learn that the lecturer is that same "Connecticut pastor" who fascinated them with the contagious imagination of "Ecce Cœlum" and "Pater Mundi." The same clear and cogent logic that in the former led us upon stepping-stones of stars to God as the father of the universe, the same glittering and brilliant style that in the latter led us through the phenomena of nature to God as the "Father of the World," is offered in "Ad Fidem" to lead us to God as our Saviour. With an air of confidence which betokens deep conviction; with an enthusiasm that is itself an evidence of Christianity, he insists upon the honest application to the Evidences of those tests which are prescribed by Christianity itself. And this is done with no juiceless language, but in a decidedly oratorical style, that will make the book very widely popular and useful. Its very fault — excess of ornamentation and gorgeousness of rhetoric — will secure a hearing for the truth by persons who might not be attracted by an ordinary book.

From the Evening Post.

We cordially thank the publishers for sending us this noble volume. It is most fittingly dedicated "To Christ and His Church." The work is full of irrefutable evidences of the Bible. In his delightful preface the learned and gifted author says, "Not only was Diderot," etc. The Typographical execution of the book is faultless.

From the New Englander.

Its merits are similar to those of his previous well known and popular volumes. The author has the gift of bold and impressive statement, ... a vivid and telling way of presentation, ... the glow and power of positive eloquence. The book will receive, as it deserves, extensive circulation, and, as we doubt not, will achieve great usefulness. We congratulate the modest and patient author upon the success which he has attained, and at which, perhaps, he himself is the most surprised.

From the Express.

The argument is strong and the style in which it is stated clear and impressive. The author is well known as one of the ablest and most interesting of the religious writers of the day.

From Harper's Monthly.

It is refreshing to come across a book written in a tone at once so candid and so cheeringly confident as "Ad Fidem." We find throughout the book, as Dr. Burr in his preface advises us we shall, "an air of great confidence." At the same time the author rarely substitutes mere assertion for argument, and never denounces as criminal the reader who fails to appreciate the force of his statements, and to accept the opinions to which they lead.

From the Princeton Review.

In this beautiful volume Dr. Burr expatiates in his favorite field of Apologetics with vigor, tact, and eloquence. He rightly traces the fortress of unbelief in the intellect to perverseness in the will and heart; shows that the difficulties in the way of religious belief are no more formidable than men encounter in every sphere of life without being stumbled by them; that with like candor applied to Christian truth all their embarrassments will vanish, etc.

From the Christian Quarterly.

These lectures discuss some of the living questions of the age in a manner at once able, pleasing, and practical. But we need not say this to those who have read *Ecce Cœlum* and *Pater Mundi.* These will know that it is almost impossible for Dr. Burr to write a dull book. *Ad Fidem* will *add* to the author's reputation. It is emphatically a book for the times; and is one of the finest defenses of the Christian religion that has been made in this country. It does for the Evidences of Christianity what *Ecce Cœlum* does for astronomy.

From the Baptist Quarterly.

This is a new work by an author who has achieved a popularity as wide-spread as it is merited. Dr. Burr writes in a style of singular freshness and vigor, groups his truths with great power, and communicates his enthusiastic earnestness to his reader.

From Scribner's Monthly.

The Rev. Dr. Burr, of Lyme, Conn., has made a sudden reputation of late by two attractive — perhaps we might even say brilliant — books on the Evidences of Christianity. He has just published a third. *Ad Fidem* is a rapid, popular, and eloquent summary of the arguments for the Bible. It is founded on careful research, and is believed to represent the latest developments of Biblical scholarship. There is no pretense of originality or appearance of scholastic learning; but the author has what is much better for his purpose, a forcible style, a dexterity in the use of striking figures and examples, and a remarkable gift of seizing and retaining the interest of his readers. He is clear, earnest, rapid, vigorous, and, above all, entertaining.

pose has been, not 'to write a book of Christian evidences merely, but to evolve the contents of the Johannean writings, which, clearly apprehended, are their own evidence, and prove Christianity itself a gift from above, and not a human discovery.' "

From the Watchman and Reflector.

"It has been known for some time to not a small circle of religious scholars that Dr. Edmund H. Sears was busy in preparing a volume on the fourth Gospel. The appearance of the book has been looked forward to with largest interest. Dr. Sears is a prominent clergyman in the Unitarian denomination. He has won a distinguished name among men of every religious faith as a profound and reverent student and a forceful, eloquent writer. The subject before him is one which has challenged the best scholarship and the sharpest criticism of the last half-century. It is a subject, too, which, like the Lydian stone, tests the quality of whatever comes in contact with it. It would not be possible for Dr. Sears, or any other thoughtful man, to discuss the Johannean writings without disclosing to his readers every shadow and shade of his theological creed, as respects all those facts and doctrines which the Church still regards as fundamental. These we give as some of the reasons which account for the wide-spread interest which has gone before the publication of this volume. And now that the book has appeared, and is being largely read, this feeling of curiosity deepens into one of great surprise. Prepared as many readers may have been for an exhibition of a most reverential spirit in Dr. Sears' studies of this Gospel, and a statement of theological views in entire opposition to the humanitarian or pantheistic theories of very many in his denomination, they were not prepared for such interpretations of Scripture, and such methods of reasoning, and such positive conclusions as fairly place Dr. Sears in the rank of orthodox thinkers and believers."

From the Cincinnati Times and Chronicle.

"Noyes, Holmes, & Co. publish a religious work by Rev. E. H. Sears that ought to enlist the attention of a very large circle of thoughtful readers. It is entitled, "The Fourth Gospel the Heart of Christ," and is devoted to elucidating the proofs of the genuineness of John's Gospel, its harmony with the other gospels, its theology, and the special light it throws on the nature, character, teachings, and works of Christ. Mr. Sears is one of the most vigorous and acute thinkers and dialecticians in America, and this stout book of 550 pages is one of the most important volumes yet contributed to theological literature in this country. It is written from a clear head and full heart; it is not dry argument or skeleton theology, but the thought glows with life, and the rhetoric is as grand and beautiful as the logic is strong. It would be folly to assume that the book has no vulnerable points which theological criticism can find ; but it is a very *vital* book, and merits the careful study of all religious readers."

From the New York Bulletin.

"Two religious or semi-theological books have just been published, which are much above the average class of literature, namely, one is, '*The Fourth Gospel the Heart of Christ*,' by Rev. Edmund H. Sears, a book of real ability, admirable spirit, and conclusive argument ; the author evolves the contents of the Johannean writings, which, he claims, clearly apprehended, are their own evidence, and prove Christianity itself a gift direct from above, and not a human discovery. Mr. Sears is on the extreme evangelical wing of Unitarianism, and his book must make a sensible impression upon thinking minds, whether they are merely intellectual, or intellectual and religious ; the two qualities are not always found in company ! "

From the New York Tribune.

"His volume will take a high rank among the biographies of Jesus which within a few years past have so greatly enriched the religious literature of the country."

From the Congregationalist.

"*The Fourth Gospel the Heart of Christ* is a book of extraordinary interest for its own sake, and still more from the position of the author, the Rev. Dr. Edmund H. Sears, as a representative of what is called Evangelical Unitarianism. Judged as a volume on its own merits, it is a rich and fresh contribution to the literature of the ages touching the life of our Lord. It is instructive and suggestive in the highest ranges of Christian thought and feeling. The title is less comprehensive than the contents of the treatise. This is not limited to the Gospel of St. John, but covers nearly all the New Testament writings, so far as they throw light on the central and controlling truth of the Supreme Deity of our Lord Jesus Christ. While establishing the genuineness and authenticity of St. John's Gospel, as tributary to the argument, the authorship of the other three gospels is established, the scope, purpose, and spirit of the book of Revelation illustrated, and the character of the Epistles of Paul largely discussed. The scholarship seems to us as accurate as it is ample. The results of wide research and critical investigation are condensed into a few pages with a clearness of statement not often equaled. The brief chapter on Gnosticism, for instance, gives a better notion of that confused and confusing mysticism than can be gathered from many columns."

From the Light of Home.

"It would be a pity that the mass of readers should be repelled from this remarkable book by its title, which suggests dogmatic controversy or textual exposition; for it is one of the most deeply interesting volumes of this generation. It is as much superior to "Ecce Homo" in power of statement, grasp of thought, and freshness of conception as that was to the Christologies of average writers. Here are the results of twenty years' faithful research and ripening scholarship. Probably it is the last, as it certainly is the best, book from the mind and heart that gave us 'Athanasia' and 'Regeneration.' With no decrease of the vigor apparent in those earlier works, there is in this the same affluence of style, and a more comprehensive reach of thought. We earnestly commend it to our readers."

From the Church and State.

"No book of recent American theology is likely to win more notice from thoughtful readers than this handsome volume by Edmund H. Sears, of 551 pages. As a work of literary art, it has great merit, and its clear, rich, and vivid style carries in its flow great wealth of thought and learning with cumulative power to the end.

"Many things may and will be said of this noble piece of thought and expression, but we choose to treat it now in its most obvious relation to our time, as a book for our age and country, and perhaps as preëminently indicating the mind of thoughtful and devout scholars of the Cambridge School. The writer reminds us often of Dr. Bushnell, and, like him, is eager to mediate between the new rationalism and the Old Gospel, yet has more substance in his thought than the eloquent Hartford divine, and is less in danger of allowing the objective reality of the Christian religion to evaporate into the volatile ether of his idealism. Mr. Sears, too, although he does not recognize duly the nature and power of the historical church, seems to come nearer to it than Dr. Bushnell, and he regards the being and mission, the death and resurrection of our Lord, more as central facts and powers of the kingdom of God with men, and less as having merely a subjective significance which is to be interpreted and applied by each individual. Both of them fall short of true catholicity in the estimate of church institutions, but Mr. Sears comes nearer the true catholic idea, and he has only to carry out what he says of the Incarnation and Atonement and the Holy Spirit, to be a good Catholic Churchman of the liberal school."

From the Literary World.

"This is a very strong book — the work of a powerful and independent thinker; and as an exposition of the Johannean theology, it has probably never been surpassed in acumen and thoroughness"

NOTICES.

From the Literary World.

"'THE ART OF PRINTING.' — It is not a little remarkable that the popular knowledge touching an event of such transcendant importance as the invention of printing should be so vague and unsatisfactory. The general notion obtains that one Gutenberg, a German, about 1450, discovered a means of taking impressions from movable type; but the particulars of the discovery are known to but few, and in the minds of many there is a doubt whether the honor of the invention belongs to Gutenberg, to Faust, or to Schœffer. This book undertakes to settle this question, and to furnish all attainable information pertaining thereto. The author has wisely chosen to give her work the shape of an autobiography, for 'The Life of Gutenberg' would attract many readers who would recoil from the prosy intimations of a 'History of Printing.' This autobiography is very pleasant reading; the little love-romance which it embodies, agreeably relieves the somewhat sombre story of the inventor's trials and misfortunes. In the preparation of her work, the author has consulted the most trustworthy authorities, and has in no instance, we believe, sacrificed the truth of history in behalf of effect.

. . . .

"The progress in the art of printing, so far as Gutenberg was concerned with it; what his partners and successors achieved, and the earliest history of printing in other lands, may be learned from the book under notice. The closing pages of this volume are occupied by a minute, accurate, and exceedingly interesting account of printing as it is done to-day, when a press can throw off 20,000 to 30,000 impressions per hour. We confidently commend this account, as clear and comprehensive, to all who are curious as to the mechanical process of book and newspaper-making. The contrast between the old and the new, the beginning and — shall we say? the perfection of the art of printing, is very strikingly presented. The author has done her work well; and hereafter there will be no excuse for the prevailing ignorance as to this interesting subject. Although her history is not exhaustive, it informs us upon all essential points, and in the pathetic story of Gutenberg's life, reveals the birth and growth of the 'art preservative of arts,' in an impressive and agreeable manner. The volume is a beautiful specimen of book-making, printed on tinted paper, with an illuminated title-page, and profusely illustrated with cuts of old and new printing implements and machinery. Altogether, in contents and externals, it is very creditable to its publishers."

From SAMUEL BURNHAM, *Editor of the Congregational Quarterly.*

"In brief, the work is interesting, emphatically instructive, well-written, and on a fresh and important theme. The writer could scarcely have hit upon a topic more attractive. A popular work, embodying the main facts of the history of printing, has been greatly desired; and in our opinion, this book meets that want. It has the rare merit of being entertaining as a story, while adhering closely to fact. It is greatly in its favor, that it has no rival in its subject."

From REV. H. M. DEXTER, D. D., *Editor of the Congregationalist.*

"I have been greatly interested in the sheets of the volume, and wonder that no book of the sort has ever before been written. Surely it cannot fail to find a multitude of interested and instructed readers, who will rejoice with me that it has been put into a shape of beauty so fitting to such a subject."

From C. HENRY ST. JOHN, *Assistant Editor of Zion's Herald.*

"'Gutenburg, and the Art of Printing' is certainly a great success, and must prove as interesting, instructing, and attractive to the general reader as to the more scientific, or those in pursuit of curious information. While the printing and illustrating will meet with due appreciation, the labor bestowed on it never can. It is the handsomest book we have received for many a day, and worthy the fame of Riverside."

From the Boston Evening Transcript.

"'Gutenberg, and the Art of Printing,' is the title of a dainty and elegant 12mo, by Emily C. Pearson, to be published in a few days by Noyes, Holmes & Co. The beautiful title-page and the historical and other illustrations add to the attractions of the carefully prepared narrative of the chevalier and artisan who brought to light the 'art of arts.' Its descriptions of the past are not, however, the only valuable portions of the book. Added to and connected with the biography is a large amount of useful information for those not familiar with the working of the material instrumentalities which belong to the measureless influence of the press."

From the Lawrence American.

"We cordially welcome this entertaining and valuable work, which is not less fascinating than the 'Schönberg-Cotta Family.' A higher than romantic interest invests the story."

"The whole story is admirably told. . . . . One of the most charming books we have read for a long time." — *Cleveland Leader.*

"Gutenberg was a hero, in his way, and his romantic story is admirably told." — *The Pacific.*

"Clear, vivid, and accurate, greatly aided by the excellent illustrations which form a marked attraction of the book." — *Detroit Tribune.*

"What might be dull history is made to partake of the interesting nature of a novel." — *N. H. Palladium.*

"Original and very attractive." — *Plaindealer.*

"It is a wonderful tale, with the advantage of all being true. The whole work is interesting." — *N. Y. Journal of Commerce.*

"The only book of the kind, and very attractive." — *Rutland Independent.*

"An exquisite volume; perfect in every appointment." — *Northern Christian Advocate.*

"A clear account of the wonderful art, and very interesting." — *Book Worm.*

"Reads like a graceful romance. A thoroughly attractive book." — *Rural New Yorker.*

"A very interesting volume." — *Philadelphia Ledger.*

"The story of one of the world's great inventors, marked by striking excellencies." — *Free Press.*

"Graceful, sprightly, effective. The whole book is exceedingly interesting, and will find a wide circle of readers." — *Interior, Chicago.*

"A beautifully printed and illustrated history of 'the art preservative,' with much useful and attractively-presented information." — *Alta Californian.*

"A book of great interest." — *Norwich Bulletin.*

"Cannot fail to please and instruct. In its subject it stands without a rival." — *Orient.*

"Of wonderful interest. A desirable acquisition to any library." — *City Item, Philadelphia.*

"Sounds like veritable romance, but has in it marvelous truth. It is worth one's while to read it." — *Amer. Rural Home.*

"Many curious facts related in an agreeable manner." — *North American, Philadelphia.*

"A most valuable book. . . . . A history of a most wonderful man." — *Nat. Temp. Advocate.*

"Very instructive and entertaining." — *M. Press, Albany.*

"It reads almost like a romance while it keeps closely to historical facts." — *Journal of Chemistry.*

"The facts, as gracefully grouped in this volume, form a most pleasing story." — *Amer. Churchman.*

"The best presentation that has yet been made of Gutenberg's work as an inventor, for popular reading." — *Lib. Christian.*

"Full of romantic interest." — *Albany Argus.*

"A valuable book, very readable, and will interest all classes." — *Hartford Courant.*

"A very attractive volume." — *San Francisco Bulletin.*

"It is curious that no such book as this has appeared in our language before. The author has told the romantic strange story in a very interesting way." — *Windham Co. Transcript.*

"An elegant sample of the art it treats of. It is the only book of the kind, affording information interesting to every one." — *Christian World.*

www.ingramcontent.com/pod-product-compliance
Lightning Source LLC
LaVergne TN
LVHW010203110826
845151LV00002B/593

* 9 7 8 1 4 2 5 5 3 5 3 7 7 *